Additives in Water-borne Coatings

Additives in Water-borne Coatings

Edited by

Gerry Davison and Bruce Lane
Consultants

RS•C

advancing the chemical sciences

The Proceedings of the meeting Additives in Water-borne Coatings – The Way Forward held on 4 September 2002 in Runcorn.

ISBN 0-85404-613-5

A catalogue record for this book is available from the British Library

Published by The Royal Society of Chemistry,
Thomas Graham House, Science Park, Milton Road,
Cambridge CB4 0WF, UK
Registered Charity No. 207890

For further information see our web site at www.rsc.org

Printed by Athenaeum Press Ltd, Gateshead, Tyne and Wear, UK

Preface

This book derives from presentations made at an International one-day conference organised by the Speciality Chemicals Sector of the Industrial Division of the Royal Society of Chemistry, and by the Paint Research Association, on 4th September 2002 held at The Heath, Runcorn, Cheshire, UK. The conference follows on four years after the previous RSC conference on this topic and updates on this advancing field with the recent developments. It is the fourth conference in the series, the previous conferences being published as

Advances in Additives for Water-based Coatings published in 1999 (RSC Special Publication No 243, ISBN 0-85404-754-9)

Waterborne Coatings and Additives published in 1995 (RSC Special Publication No 165, ISBN 0-85404-740-9)

Additives for Water-based Coatings published in 1989 (RSC Special Publication No 76, ISBN 0-85186-607-7)

It would not be possible in a one-day conference to cover the whole field of additives for water-borne coatings. The approach taken, both in the meeting programme and in the edited papers in this publication, was to seek two plenary lectures from world authorities to overview the field and its direction, and to select key additives currently employed for in-depth treatment of their use, behaviour and scope by expert practitioners in those additives. Particular emphasis throughout is placed on the practical aspects of the subject, theoretical considerations only being introduced where essential in order to more fully understand the topic. As the meeting title implied, the programme was essentially a ' look forward' rather than a 'review' of existing practice. The developments discussed are considered, most importantly, in the light of increasingly stringent environmental requirements and a background of regularly changing legislation.

The plenary papers comprise Chapters 1 and 5. In Chapter 1, Roger Brown [Akzo Nobel Decorative Coatings] describes how the use of certain additives, which otherwise on cost and general property grounds one would like to avoid, become essential to impart the desired properties to the coating system. In Chapter 5, Jon Graystone [Paint Research Association] presents a means of how, by assessing the likely performance of an additive in a coating using a number of 'Additive Rules', optimisation of properties may achieved.

Specialist papers on particular additive types comprise the remainder of the publication. Two chapters are devoted to Rheological additives, reflecting the importance of rheology control in water-borne coatings. In Chapter 3, Dirk Kruythoff [Hercules BV] reviews how the use of Cellulosic rheology additives has developed in recent years to overcome shortcomings in these materials and to enhance their performance. In Chapter 6, Isabelle Mussard [Rohm and Haas] covers recent advances in the field of Associative Thickeners which attempt to broaden the range of rheological control these polymers can impart.

v

One chapter each is devoted to Biocides, Dispersants, Matting Agents, Antifoams and Surface Active Additives. In Chapter 2, Scott Betts [Thor Chemicals] considers how a wide range of water-based coatings are susceptible to attack, leading to, for example 'In-Can' and 'Dry Film' damage by a number of different organisms, and explores the range of Biocides which can be used to address these issues. Ian Maxwell [Avecia] in Chapter 4 addresses Dispersants, and shows how improved additive design can help formulators address difficult water-borne coatings formulation requirements in respect of increased black jetness, transparent iron oxide formulations and flake alignment at higher solids contents.

Chapter 7 from Hans-Dieter Christian [Degussa AG] emphasises that surface appearance can be an important requirement in water-based coatings, for instance, in the continuing trend, particularly in the automotive and furniture industries, for matt surfaces. This has led to a new generation of untreated silicas to bring about new standards in matting efficiency. In the penultimate Chapter 8, Hans-Guenther Schulte [Cognis] considers the subject of Antifoams. He starts by describing the mechanisms of foam generation and control in some detail and then goes on to show how a range of de-foamers including mineral oils, emulsions, powders and silicone-based types may be employed. The various methods of assessing the performance of a de-foamer are described. Finally, in Chapter 9, Donna Perry [Dow Corning] indicates how surface active agents, most particularly silicones, are used to reduce the surface energy of a coating to which they are added, in order to achieve certain surface properties including desired flow, levelling and wetting performance.

The editors hope that this publication will prove to be a valuable addition to current literature on this topic, suitable for both readers relatively new to the field as well as experienced workers in the coatings industry.

Gerry Davison and Bruce Lane
March 2003

Contents

ADDITIVES IN COATINGS - A NECESSARY EVIL?

Roger F G Brown

Pigment Dispersion Group, Technology Centre Decorative Coatings
Akzo Nobel Decorative Coatings
PO Box 37, Crown House
Hollins Road, Darwen, Lancashire, BB3 0BG, UK.

1   INTRODUCTION

It is generally accepted that surface coatings are applied to decorate and protect (paints and lacquers) or to inform (paints and inks). Coatings in their simplest form as a dried film on the substrate consist of binder (lacquers) or binder and pigment (lacquers, paints and inks). Additionally, it is necessary for the paint to be applied to the substrate. With the exception of some curable liquid formulations (e.g. radiation cured, liquid two pack paints, etc.) some form of diluent is required to enable the coating to be applied. In the majority of cases, this diluent is some kind of liquid volatile material (organic or aqueous) though in the case of powder coatings the diluent enabling application is air. Hence, to generalise, liquid paint, in its simplest form, will consist of pigment, binder and diluent. In this simplified formulation, the only variables are pigment volume concentration (PVC) and solids content (SC) and it is straightforward to record any given measurable property of the wet or dry paint as a function of the composition.

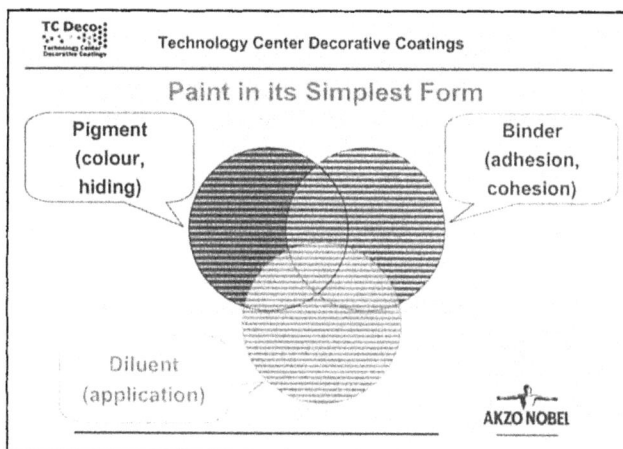

Figure 1 *Paint in its Simplest Form*

An optimised formulation with the best balance of properties can then be readily determined. Unfortunately, in the vast majority of, if not all cases, this optimised balance point is deficient in a number of key areas and, in some areas, the required level of performance simply cannot be achieved in such a simplified formulation. This is where additives play a part. By selective use of additives, it is possible to achieve better properties in specific performance attributes that cannot be achieved with the simplified tertiary formulation. Additionally, it becomes possible to achieve an enhanced balance of properties, where some properties can be improved without the same degree of detrimental effect on other properties. Additives are used in coatings for a great number of purposes to improve specific properties throughout the life-cycle of a paint such as processing properties in manufacture, in-can stability, application properties, drying properties and final film performance. Unfortunately, each extra additive adds a degree of complexity to the formulation and achieving the optimum balance point between varying (often opposed) properties. Hence, from the paint developer's point of view, a minimalist approach to formulation is to be preferred. Nevertheless, because it is generally impossible to achieve the desired properties without additives, this increase in formulation complexity is justifiable and wholly necessary. With VOC legislation stimulating the development of alternative technologies such as waterborne, high solids, radcure and powder, the need for additives will not go away. If anything, with waterborne technologies supplanting solventborne in a number of sectors, it can be argued that the importance of additives is increasing due to the added complexity of formulating in water.

2   LEGISLATION

It is not intended to provide a comprehensive list of legislation relevant to the coatings industry. However, within the European market, a number of legislative acts have been passed with many subsequent amendments and adaptations to technical process which have great significance for the Coatings industry. A number of important Directives are presented in Table 1. These have profound implications for the formulator in terms of labelling, inclusion of specific products and the levels at which specific products may be used. Furthermore, there are a number of voluntary labelling initiatives within the Deco sector such as Ecolabel, Blue Angel and the Globe. Additional to this, there are also sector specific directives currently under consultation, such as the Deco Directive on VOC's, which will further add to the framework within which the formulator must work. In many cases, it is this framework which drives product development but also the conversion from one type of technology to another. Environmental issues, both legislative and voluntary, will continue to drive additive development and paint development simultaneously for a long time to come.

Table 1   *Selected legislation of relevance to the coatings industry.*

Dangerous Substances Directive (DSD)
Dangerous Preparations Directive (DPD)
Marketing and Use Directive (MUD)
Solvent Control Directive (SCD)
Biocidal Products Directive (BPD)

3   THE NEED FOR ADDITIVES

As already discussed in the introduction, additives are frequently essential to obtain properties that cannot be obtained from the simple combination of pigment, binder and diluent. Additives are multifarious in nature and are used to control a number of different properties. A list of properties that may be controlled by selective use of additives, by no means exhaustive, is presented in Table 2, along with examples of the chemical nature of such materials. Paint formulation is a balancing act. This is because paint is expected to fulfil a number of key requirements, which will vary, dependent on the application. To maximise profitability, paints should be cheap to manufacture both in terms of raw material costs but also ease of production with high throughput, low energy processes preferred. Technically, paints must have good in-can stability so that there is no separation of ingredients or colour. Paints are required to have good application properties, specifically associated with rheology, whether for brush application, spray application, dip coating, coil coating, etc. dependent on the sector. The paint should dry well with respect to coalescence or cure. Finally, the final paint should adhere well to the substrate, have the correct colour and level of gloss, show good retention of colour and gloss and have the desired protective properties, ideally for as long possible. Hence, there are a great number of properties that a paint is expected to have, some of which are specifically opposed (e.g. sag resistance vs. levelling) and there is no such thing as the perfect paint. It is obvious that a balance must be struck between opposed properties. One of the key requirements when formulating a paint is to establish what balance of properties should be achieved. Within this, tacit acceptance must be made that the perfect paint does not exist and that the paint may not be able to match all desired properties. In establishing the desired balance, the most important properties for the specific application must be defined so that, where a trade between properties is required, the relative weighting given to gains and losses reflects their relative importance. Less obvious to non formulator's, is that non opposed properties also need to be balanced because each additive, as well as affecting desired properties, coincidentally will usually have knock-on effects, often unexpected, on other properties. For example, drying agents added to a paint to improve curing properties may migrate to the surface of the film and have a detrimental effect on gloss and/or gloss retention. However, this is where the skill of the formulator comes to the fore, and paint companies are able to differentiate their own products in terms of the performance balance achieved for a specified price. The type and quantity of each additive used must be carefully selected, based on experience, and optimised, based on well designed experiments, to achieve the best balance of properties, taking into account the relative importance of each property. Hence, whilst formulation is easier with fewer components, the increasing complexity of including multiple additives allows each paint manufacturer the chance to make their products distinguishable in the market. However, when increasing the number of additives used, care should be taken that particular combinations are not antagonistic. The ideal situation, obviously, is where synergy is achieved within the additive cocktail used.

Table 2   *Typical functional modifiers used in the coatings industry.*

Function	Typical chemical nature
Wetting, dispersing, stabilising	Numerous anionic, non-ionic, cationic and amphoteric surfactants/dispersants
Anti-foams	Hydrocarbons, silica, silicones
Biocides, fungicides	Thiazolinones, oxazolodimes
Rheology modification	Clays, amide modified alkyds, hydrophobically associating water soluble polymers, cellulosics, titanium chelates
pH adjustment	Amines, amino hydroxy compounds
Anti-skin	Mekoximes
Open time/lapping	Glycols, glycol ethers, thickeners
Adhesion promotion	Silanes
Flash rust inhibition, anti-corrosion	Sodium nitrite, phosphates
Film formation	Glycol ethers
Drying	Metal soaps, benzophenones
Matting agents	Silica, waxes
Slip agents, block resistance	Silicone waxes
Durability	Triazines, benztriazoles

4   WATERBORNE vs. SOLVENTBORNE – INCREASING THE NEED FOR ADDITIVES

There are a number of fundamental differences between solventborne and waterborne paints. When comparing, for example, a latex trim paint with its solventborne counterpart, there are a number of features that have been taken into account when developing the two formulations. Firstly, the high surface tension of water must be considered. This makes pigments more difficult to wet and stabilise and makes substrate wetting more difficult and generates a requirement for the use of surface active additives to help overcome this. Secondly, the disperse nature of the binder should be considered. This has direct effects in that surface active material is usually needed to stabilise the binder but also has indirect effects such as rapid increase in viscosity with increasing solids content and the need for phase inversion and colalescence during the film formation stage. There is a further complication with increased crowding in a latex system but this is considered outside the scope of this paper.

In a dispersion (waterborne) paint, viscosity is generally very low (reflecting the viscosity of the medium) below a volume fraction of particles of 40% whilst it goes up increasingly rapidly above this value. Additionally, because the binder is not soluble in the medium, once the binder particles come into contact, the film cannot be spread further by working in extra medium. This leads to a generalised phenomenon in waterborne paints whereby open time is short. Because the viscosity can increase very rapidly with increase in solids content due to evaporation of a small amount of water, becoming irreversible at a certain point, it is necessary to formulate at sufficiently low volume solids to obtain satisfactory open time. Formulating at low volume solids leads to paints of low viscosity and thickening agents are required to give adequate in-can viscosity and film build (body) to the paint.

By comparison, in a solution (solventborne) paint, the viscosity increases in a much more controlled manner as the volume solids increases during solvent evaporation and it is also possible to work extra solvent into the existing film to thin the paint back during application. Hence, it is possible to formulate with good in-can viscosity and film build, whilst still retaining sufficient open time, levelling and flow, without the need for specific thickening agents.

Also, when comparing dispersion and solution coatings, in the latter, the binder is already part of the continuous phase in the wet paint whilst in the former, the binder is dispersed in the continuous phase. Hence, for a solution coating, the continuous phase becomes gradually more concentrated during drying until all of the solvent has evaporated. Conversely, for a dispersion coating, there is a point of phase inversion where the binder particles touch and deform and change from being part of the dispersed phase to being the continuous phase. Following this phase inversion, to develop film integrity, it is important that the binder particles deform and coalesce to fill the voids left by the still evaporating medium and coalescing solvents are frequently needed to assist in this. Hence, it can be seen that many properties that can be taken for granted with solventborne paints must be specifically provided through use of additives in waterborne paints.

To summarise the comparison between alkyd trim paints and their waterborne counterparts, a lot of properties are provided by solventborne alkyd technology for which specific additives are required in waterborne dispersion paints. Frequently, the polar alkyd resins are capable of wetting and stabilising the pigment for which dispersing agents are needed in waterborne.

Additionally, during drying, solventborne alkyd paints have tendency to spontaneously stratify to yield a thin resin rich layer at the air interface. This gives high gloss, further reducing the need for specific wetting and dispersing additives to improve the degree of dispersion of the pigment, necessary in waterborne paints to achieve high gloss. The relationship between volume solids and viscosity is frequently such with alkyd paints that good in-can viscosity and film build can be obtained at the same time as good open time, levelling and flow, without the need for specific thickening agents, as needed in waterborne. The fact that the resin is already present as a constituent in the continuous phase in solventborne paints alleviates the need for coalescing solvents which are frequently needed to promote development of film integrity in waterborne paints following phase inversion. Finally, because of the aqueous environment, there is generally a need for biocidal additives in waterborne paints, which are frequently not required in solventborne paints. Therefore, it can be seen that changing from solventborne to waterborne technology is likely to entail an increase in the requirement to include additives in the formulation.

5   SUMMARY

From a simplistic point of view, the formulator's job is easiest with the minimum number of paint ingredients. A response surface can be easily constructed for a four component system by use of an equal dimensioned tetrahedron.

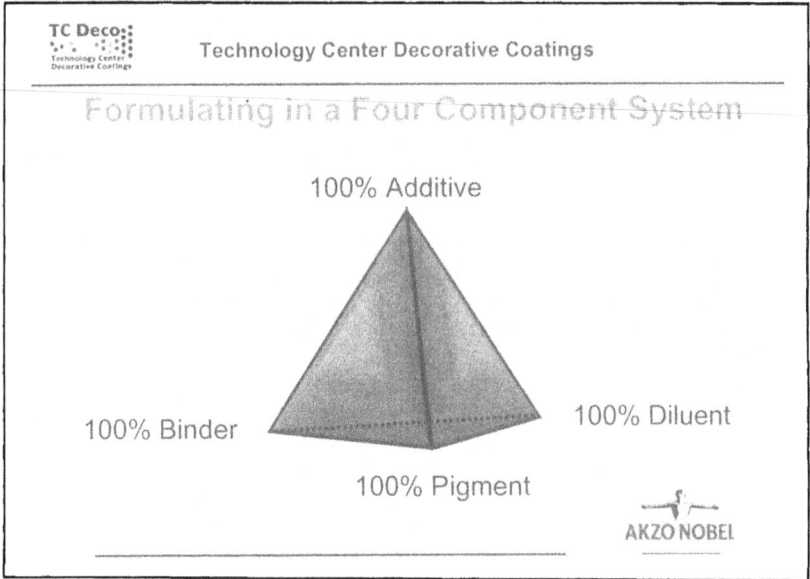

Figure 2 *Formulating in a Four Component System*

This would allow for the ratio of pigment, binder, diluent and a single additive/performance modifier to be easily optimised. Provided that the correct weighting is assigned to each measurable property, commensurate with its importance, the best balance of performance is readily achievable. However, because of the simplicity of achieving this, it would be extremely difficult for individual paint manufacturers to differentiate their products in the market unless they have proprietary materials available with superior performance. Fortunately/unfortunately (dependent on the viewpoint) paint formulation is not this simple and it is generally not possible to achieve a satisfactory balance of properties without a multi-component formulation containing a range of different additives. This is where the skill and experience of the paint formulator to maximise synergy and minimise antagonism allows products to be differentiated in the market. Hence, additives in coatings are an integral and necessary part of paint formulation to give enhanced performance. With environmental concerns driving formulation development and a gradual move from conventional solventborne technology to technologies with reduced environmental impact, the need for additives in coatings is set to increase. Although inclusion of a number of different additives increases the complexity of the formulation, additives are fundamental in improving the balance of performance and providing market opportunities – truly a necessary evil.

BIOCIDE REVIEW

Scott Betts

Thor Specialities (UK) Limited, Earl Road, Cheadle Hulme, Cheshire SK8 6QP, UK

1 INTRODUCTION

A wide range of water based products, including paints, polymer emulsions, building products and the like are usually susceptible to attack by a number of different micro-organisms. Such attacks can result in severe economic loss and it is, therefore vital to institute measures that will prevent this from happening. The use of Biocides has been extremely important to thwart economic damage caused to paints and other industrial products. Microbial damage in coatings can be categorised into two types: In-can Damage and Dry Film Damage.

2 IN-CAN DAMAGE: INFECTION OF INDUSTRIAL PRODUCTS IN THE WET-STATE

If a coating is not protected with the right biocide, microbial damage could occur. Damage to paints or other systems can manifest itself in a number of different ways.

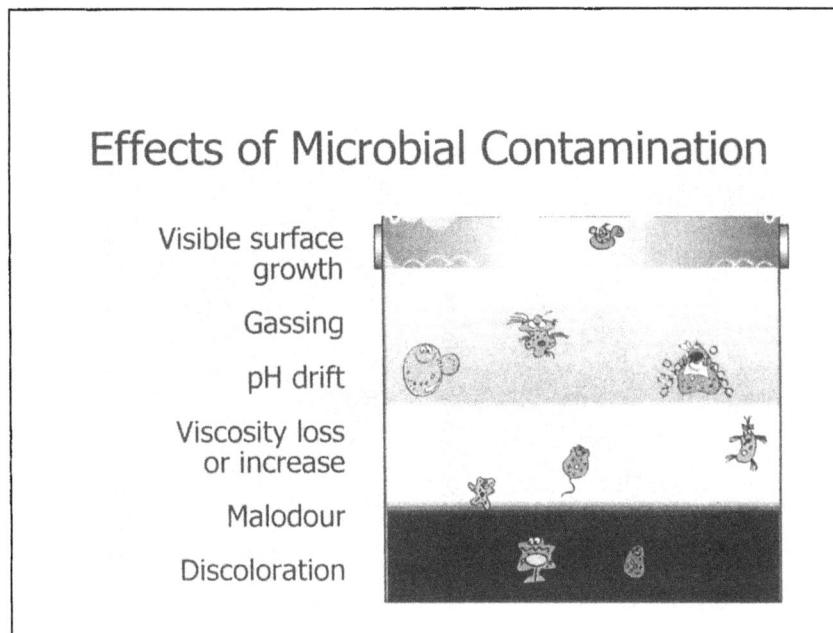

Effects of Microbial Contamination

Visible surface growth

Gassing

pH drift

Viscosity loss or increase

Malodour

Discoloration

Figure 1 *Effects of Microbial Contamination*

Firstly, there is the possibility that visible surface microbiological growth could occur. Here the surface of the product is visibly fouled. Fungal growth can be seen as usually green, black or off-white growths on the surface. Occasionally pools of bacterial colonies are seen. These appear as off-coloured product spots on the surface.

Secondly, damage can been as product separation. Coatings contain varying amounts of solids depending on its function. To keep these solids suspended in a water medium, other additives such as cellulosic thickeners and LAS surfactants are used. These components can be attacked by microorganisms and can be broken down. As this happens, their ability to function properly diminishes and the solid matter of the formulation settles to the bottom of the container.

Thirdly, depending on the type of microbial infection, gas production is known to occur. Gases such as carbon dioxide are generated as the result of microbial activity. As gas is produced, pressure can build up in the package causing swelling. In rare cases, the pressure has been sufficient to cause the lids of the container to buckle and is forced off the container. Other gases such as hydrogen sulphide can be given off if anaerobic bacteria infect the product. In this case the odour of "rotten eggs" is perceptible.

Fourthly, microbial infections can produce unpleasant odours as microbial by-products build up. Production of hydrogen sulphide (as described above) and various organic acids can result providing the noticeable odours.

Other problems of product discoloration, pH drift, loss of viscosity, loss of performance can also result from products contaminated with microorganisms. No matter how the infection manifests itself, the consumer does not want to purchase a product that has the above issues.

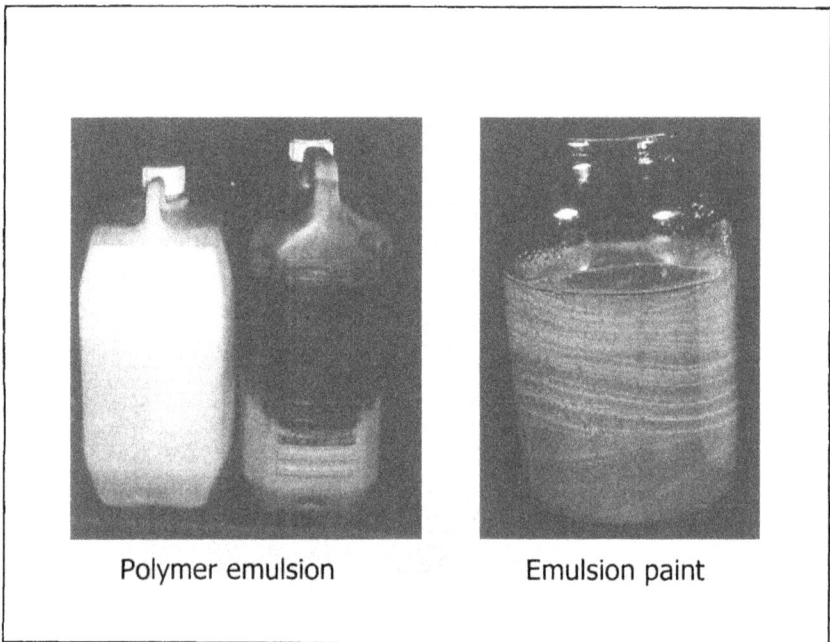

Polymer emulsion   Emulsion paint

Figure 2 *Unattractive Products*

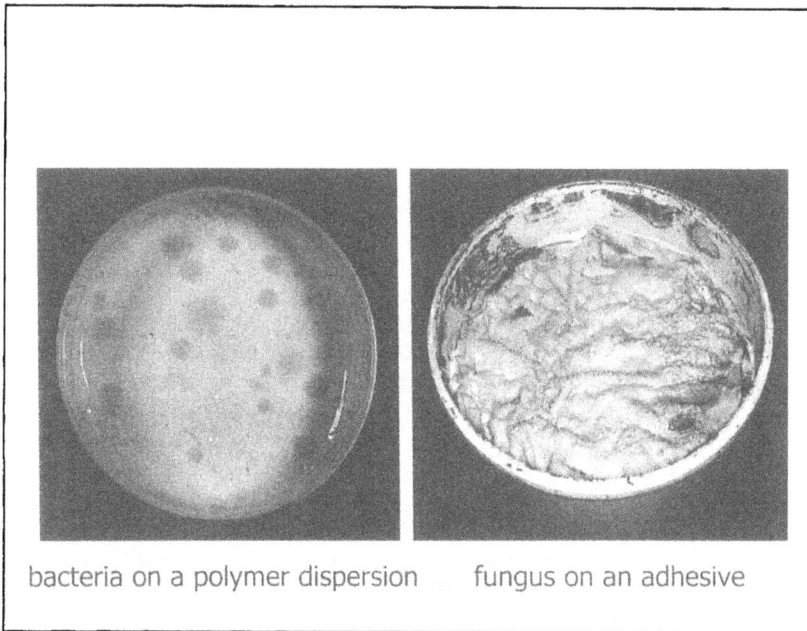

bacteria on a polymer dispersion   fungus on an adhesive

Figure 3 *Further Unattractive Products*

Many different species of microorganisms have been responsible for paints spoiling. Recently, it has been observed that the most prevalent species are from the *Pseudomonas, Burkholderia,* and *Glucanacetobacter* genera.

3 DRY FILM DAMAGE: COLONISATION OF FUNGAL/ALGAL SPECIES ON THE DRIED PAINT FILM

The second category of microbiological spoilage is related fungal or algal growth on surface coatings.

Fungi and Algae are distinctly different groups of organisms, but it is extremely difficult to distinguish the type of organism growing on a surface without the aid of a microscope. This has resulted in mis-identification of the causative organism in many instances. However, a simple rule is that algae will only grow in the presence of light, preferably sunlight, so are usually only found outdoors, while fungi may grow in both exterior or interior locations, but generally prefer to grow on interior surfaces.

Inside the home moisture build-up may be due to poor ventilation allowing condensation on cool surfaces, particularly in laundries, kitchens and bathrooms. Moisture levels can also rise due to leaks in weatherproofing of windows, doors or roofs, or from rising damp from faulty damp course in the foundations or exterior walls. Moisture build-up on exterior walls and structures can be from rainfall, condensation, broken gutters, overspray from garden watering systems or faulty ground drainage.

The amount of direct sunlight a surface receives will effect the drying of the surface, so in the Southern Hemisphere, south facing sides of buildings, fences, posts and other structures stay damp much longer than surfaces facing other directions. The shady side of a building or areas protected by overhangs, such as eaves, porches and carports, or areas shaded by shrubs and trees will exhibit greater amounts of surface growth due to the tendency to stay wet longer.

The texture of the surface is also important. Hard glossy surfaces will resist the accumulation of dirt and other organic matter, while textured finishes will catch and retain greater amounts of material. The accumulated material will assist to hold moisture and provide minerals and basic nutrients that organisms need for growth.

Algae

Algae are simple plants that range in size from microscopic single celled organisms to 50 metre long seaweed. They are very adept at growing on a wide variety of surfaces as they merely use them as an anchor point, needing virtually no nutrients from the surface (carbon and energy sourced from photosynthesis). Algae are commonly found growing on painted surfaces, masonry, concrete, paving, timber, and stone and in any pool of water.

Algae do not attack or directly damage the surface they are located on, but the growing algal colony will retain moisture making the surface unsightly and slippery, and also assist fungi and other higher order plants to grow. The retained moisture may also penetrate paint or other coatings, leading to their premature failing.

Typical algae found on surfaces include Chlorella, which looks green when the surface is wet, but black or blue green when the surface dries out. The other most common algae are Trentepohlia, which may show a variety of colours from dark blue-green to yellow, red, brown, black or even orange.

Fungi

Fungi differ from Algae in many important aspects. The most significant is that fungi do not produce their own food (unlike algae) and therefore have no need for sunlight. They are able to grow in the presence or absence of light and obtain nutrients necessary for growth by breaking down or decaying organic matter. These may be natural such as plant matter, or items made from plants, or synthetic compounds such as cellulose or surfactants from water based paints, or resins and oils in solvent based paints.

Fungi must take nutrients from the surface they are growing on or the surrounding environment. Fungi are commonly found growing on plant surfaces, cut timber, painted surfaces and almost anything else that contains sufficient moisture, including glass if the humidity is high enough. They prefer to grow in damp dark places. They are found on exterior paints, on eaves and ceilings of porches, carports and other open buildings where leaks or condensation or humidity allow enough moisture to build up and they are protected from direct sunlight.

Fungi are capable of penetrating paint films and breaking down some components of the paint. This may result in failure of the paint film leaving the underlying substrate open to moisture penetration and perhaps further degradation by wood destroying fungi, or rusting of steel surfaces.

Figure 4 *Fungal Growth on Timber*

Fungi also commonly grow on grout or sealant in bathrooms or showers. Fungi growing in these areas find it very difficult to extract nutrients from the grout or sealing material, and so rely on residues of soap and body fat to provide the nutrients they require.

Fungi typically found growing on surfaces in buildings include members of the *Aspergillus, Aureobasidium, Cladosporium* and *Penicillium* genera.

4 BIOCIDE USE

Since various industrial products have known spoilage problems as mentioned above, the use of Biocides is essential. Industrial preservation has revolved around the use of isothiazolone (IT) and formaldehyde donor (FD) biocides.

IT based products offer some key advantages as noted in Table 1. For in-can biocide activity, the two IT based products most commonly used are 5-chloro-2-methyl-4-isothiazolin-3-one (CIT) / 2-methyl-4-isothiazolin-3-one (MIT) (CIT/MIT are sold together in a ratio of 3 parts CIT and 1 part MIT) and 1,2-benzisothiazolin-3-one (BIT) Table 1 shows the advantages of FD type biocides. In many parts of the world, IT combined with FD biocides are very popular. They offer key advantages such as increased killing power and enhanced stability of CIT/MIT.

Table 1a *Strengths and weaknesses of wet state biocides*

Biocide	Strengths	Weaknesses
CIT/MIT ratio 3:1	Very strong antimicrobial activity Use levels in formulations 10 to 30 ppm Very cost effective Broad spectrum against fungi/ bacteria Indirect food contact approvals Not persistent in the environment Fast kill rate	Poorly stable above 40-50 °C Poorly stable above pH 9 Poorly stable with redox agents Poorly stable with 1 and 2 amines & thiol groups Skin sensitising potential Increasing regulatory pressure on CIT/MIT (>=15 ppm label limit)
BIT	Better pH, temperature, redox stability Indirect food contact approvals Use levels in formulations 50 to 200 ppm Not persistent in the environment	Efficacy gaps in Pseudomonas/ yeasts/fungi Skin sensitiser >=500 ppm label limit Less cost effective than CIT/MIT Slower kill rate
MIT/BIT	Better pH, temperature, redox stability Improved activity against Pseudomonas Indirect food contact approvals Use levels 50 to 100 ppm Lower sensitising potential than CIT/MIT Better cost performance than BIT Not persistent in the environment	Slow kill rate >=500 ppm label limit for BIT Performance gap against yeast in acidic conditions
Bronopol	Good antibacterial activity Indirect food contact approvals Not a sensitiser Acceptable environmental profile Fast kill rate Excellent in combination with isothiazolone biocides	Performance gap against yeast Poor heat and alkali stability Less cost effective than CIT/MIT
Formaldehyde Donors	Good antibacterial activity Fast kill rate Good heat and alkali stability Cost effective Excellent in combination with isothiazolone biocides	Formaldehyde itself is a sensitiser Performance gap against yeasts& fungi

Table 1b *Strengths and weaknesses of dry film biocides*

Biocide	Strengths	Weaknesses
Carbendazim	Antifungal performance Leach resistant Very cost effective Causes no colour shifts Good pH and heat stability Blends well with other biocides	Efficacy gap against Alternaria sp. Future classification as Category II Mutagen Cannot have in clear systems
OIT	Good broad spectrum fungicide No colour shift in coatings Can use in clear formulations Good stability in coatings Blends well with other biocides	More expensive than carbendazim Labelling as a sensitiser at 500 ppm Can have leaching
IPBC	Good broad spectrum fungicide Can use in clear formulations Blends well with other biocides Good stability in coatings	Can cause colour shifts in coatings More expensive than carbendazim Some leaching possible Contains halogens Newly classified as a sensitiser
ZPT	Good broad spectrum fungicide Leach resistant Blends well with other biocides Good stability in coatings No labelling issues Not a sensitiser Some algicidal activity Blends well with other biocides	Can cause colour shifts in coatings More expensive than carbendazim Cannot use in clear coatings
Diuron	Good antialgal performance Does not cause colour shifts Very cost effective Can have in clear systems Blends well with other biocides	Classified as a Category III mutagen Some leaching possible Halogen containing
Terbutyrn	Good antialgal performance Does not cause colour shifts Very cost effective Can have in clear systems Leach resistant Halogen free Blends well with other biocides	Does have an odour

These products have not come without their problems. Emulsion producers had to carefully control redox potential when using CIT/MIT based products and not add the biocide when the emulsion was too hot. The paint producers had to be careful and not to add the biocide with other very alkaline raw materials or degradation of the CIT species could occur. The industry has gone through a learning curve on how to successfully use these new organic biocides. Industry continued to master the use of IT and FD based biocide because they offered the more cost effective solution. However, new regulatory pressures have appeared.

The 28th Adaptation to Technical Progress in relation to the European Dangerous Substances Directive (67/548/EEC), has now classified CIT/MIT in any preparation equal to or greater than 15 parts per million to be a sensitiser. Since many formulations will require more than 15 ppm of CIT/MIT, it will make it more difficult to use this chemistry (and not put hazard labels arising from CIT/MIT use) and achieve technical success.

To prevent paints being classified as sensitisers, producers have opted for products that have a low CIT/MIT content boosted with a formaldehyde donor or bronopol. Such products include ACTICIDE® FS(N) and ACTICIDE® LA0614.

A further strategy to prevent this classification is to use a CIT free system. Such options could include the use of BIT/MIT. Additionally use of bronopol or FD biocides help to enhance preservation of this combination. Such products include ACTICIDE® MBS, ACTICIDE® MBL, ACTICIDE® MBF50 respectively. See Table 1 for details.

Additionally Europe is faced with implementation of the Biocidal Products Directive (The BPD-European Directive 98/8/EC). The directive requires suppliers of biocides to register biocidal formulations and active ingredients by providing an extensive data package (toxicology, environmental, chemical properties and efficacy data required). Many active ingredients will need to have millions of Euros spent to generate this data. At such a cost, many active ingredients will not survive through the implementation periods. If one considers similarly extensive biocide regulations in North America, an estimated 75% of global biocide usage now becomes highly regulated.

With so much regulatory pressure on the use of biocides for the future, the challenge will be to take the "favourable" active ingredients and produce blends that offer as much value as possible. As such, the next few years are going to be very interesting indeed!

CELLULOSE ETHER THICKENERS - KEY ELEMENTS FOR COMPLETE RHEOLOGY SOLUTIONS

Dirk Kruythoff

Hercules BV - Aqualon Division
Noordweg 9
P.O. Box 71
3330 AB Zwijndrecht
The Netherlands

1 INTRODUCTION

The use of rheology additives such as clays, plant exudates and natural polymers to formulate paints dates back to ancient times. These materials are used to thicken the fluid, suspend dispersions of additives in the fluid and improve the stability of the ensuing dispersion as a function of temperature and shear history. This paper classifies cellulosic rheology modifiers with respect to their influence on paint properties related to rheological profiles.

2 BACKGROUND SCIENCE

The science of rheology is very useful to get better understanding of the phenomena related to application characteristics. For that reason it might be beneficial to first have a closer look at some of the rheological aspects that play a role in coatings.

There are three major sources that contribute to the increase in viscosity of water-borne paints;
1. An increase of the viscosity of the continuum, which is water in this case, due to dissolved polymer
2. An increase of the hydrodynamic volume of the condensed phases, such as latex and pigment, due to adsorbed polymer
3. Enhanced interaction, which covers a variety of phenomena, such as
   - bridging of particles (surface to surface)
   - interaction of adsorbed layers with each other
   - interaction of adsorbed layers with polymers in the continuum
   - ....

What is understood by the term "traditional thickener" in general is a thickening agent which, when present in the coating, increases the viscosity of the aqueous phase and which do not significantly associate with the other ingredients. The polymer concentration at which polymer coils start to overlap is indicated by the critical polymer concentration ($c*$). Above this concentration, the viscosity is increased by a "traditional thickener" polymer dissolved in the aqueous phase, because of the entanglement of the polymer chains. (Fig. 1).

Figure 1   Mechanism Traditional Thickener
                Aqueous Viscosity vs Polymer Concentration

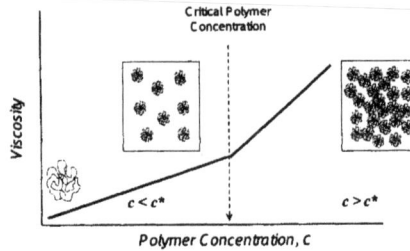

3 CELLULOSE-BASED THICKENERS

The largest group of materials of this type, which is solely dependent on viscosity development of the water phase, is the series of cellulose-based thickeners. Cellulose itself is very interesting material. Constituting the skeleton of trees and a wide variety of plants, it is a fantastic source of renewable furnish. The polymer itself contains hydroxyl groups making the material very hydrophilic. Because of hydrogen bonding effects the polymer chains are to a large extent aligned in a crystalline setting. Consequently, cellulose itself is not water-soluble. The structure, however, will either swell or break up under alkaline conditions. This breakdown of the crystallinity can be made permanent by (partly) substituting the hydroxyls, thus making a return to the original situation impossible (Fig.2). The level and type of substituent introduced largely determines the solubility parameters of the resulting cellulose derivative.

Figure 2   Cellulosic Rheology Modifiers ...*The Principle*

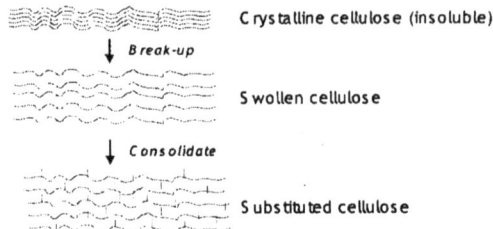

A single cellulosic polymer chain is built up by a series of anhydroglucose units, which are linked together via the beta 1,4 position. Each unit contains three hydroxyl groups, which can be made available for substitution by etherification (Fig. 3). The resulting cellulose-ether within the various substituent types is characterized by
    -the degree of polymerisation (DP), that is the number (n) of anhydroglucose units forming one polymer chain.
    -the degree of substitution (DS) per anhydroglucose unit, varying from 0 to 3.

-the molecular substitution (MS) per anhydroglucose unit, which can go beyond 3 in case the substituent itself also contains hydroxyls available for further etherification steps.

Figure 3 *The Cellulosic Polymer Chain Structure*

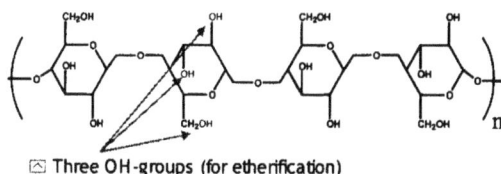

⬠ Three OH-groups (for etherification)

The following overview covers the types of cellulose ethers that are widely used in waterborne coatings.

HEC/EHEC
Substitution of the hydroxyls by hydroxyethoxyl groups, eventually combined with ethyl substituents, not only enhances the permanent non-crystallinity of the cellulose backbone, but at the same time keeps the material in a hydrophilic state. The material is therefore very suitable for the purpose and thus often used as a paint thickener.

HM(E)HEC
This group also incorporate hydrophobic agents, which make the materials very attractive in formulations leading to enhanced performance as will be explained later

MC
Substitution by methylgroups is also a means to bring about solubility. However to ensure proper solution properties under application conditions the methyl groups are usually combined with hydroxyethoxyl or hydroxypropoxyl groups.

CMC
Another substituent that is used in paint applications is carboxymethoxyl. The resulting thickener is usually available in its sodium salt form and thus exhibits an anionic character. Needless to say, this charged material is extremely hydrophilic.

4 SPECIAL CELLULOSIC THICKENERS

4.1 Hydration Retarded Cellulosics

Water solubility is an important characteristic for a polymeric material to make it a potential thickener of coatings, but the material can be further adapted to make it well suited for its application environment. Good examples are additives enhancing the hydration and dissolving properties. In order to avoid lump formation and to render sufficient time to adequately disperse the material, powdered material can be treated to retard the hydration time of the polymer without affecting the dissolution time, which is the crucial factor (Fig. 4).

Figure 4 *Treated Cellulosic Polmers to Extend Hydration Time*

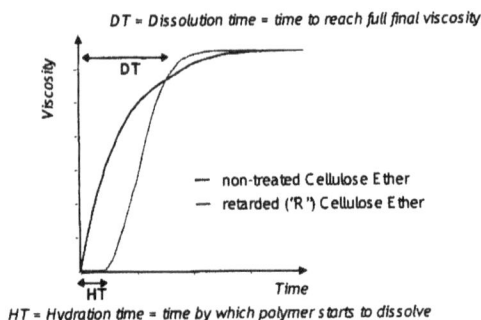

4.2 Biostable Cellulosics

Once a system has been thickened, the material should retain its rheology unchanged until it is applied. However, there are some specific species that can adversely affect the final coating over time. Problems of viscosity loss often have their origin in the very early part of the paint production. Bacteria, fungi or yeasts may be incorporated as unwanted components of the formulation, entering via raw materials, water, or through the production environment in general. In addition to the fact that this usually causes deficiencies like odour and discolouration, the system might reduce in viscosity due to degradation of the cellulose ether caused by enzymes produced by the bacteria. Usually paints are well preserved, but enzymes already present before the preservation step can be a potential threat.

So-called cellulytic enzymes are able to reduce the molecular weight once they are present. The point of enzymatic attack is those glycosidic linkages adjacent to unsubstituted anhydroglucose units. The enzymes are able to break these linkages, resulting in viscosity loss (Fig. 5).

Figure 5 *Possible Points of Enzyme Attack on Conventional Cellulosic Thickeners*

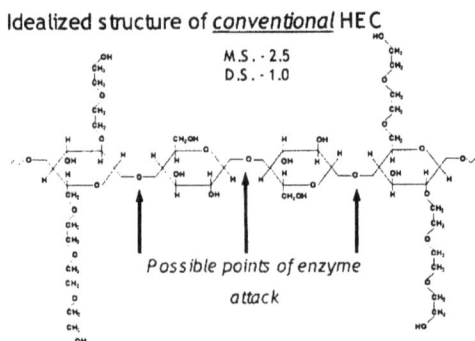

The way to increase the enzyme resistance of cellulose ethers is to limit the amount of unsubstituted anhydroglucose units by adjusting the reaction conditions accordingly. In other words, to increase the degree of substitution rather than the molecular substitution, as is shown in this Fig. 6. The resulting cellulose ether is known as biostable or "B"-grade material.

Figure 6

Idealized structure of *HEC B-grades*
HEC with increased enzyme resistance)

M.S. - 2.5
D.S. - 1.7

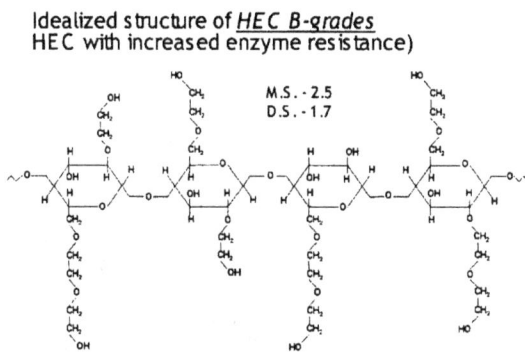

5 PSEUDOPLASTICITY WITH CELLULOSIC THICKENERS

MC, (E)HEC, and CMC Cellulosics all introduce pseudoplastic behaviour to the paint. However, the extent of the effect differs. This can be directly linked to the state of solvation for the different rheology modifiers. A good example is CMC, which is very hydrophilic and the molecules are thus expected to be generally more stretched or uncoiled once dissolved. The effect of shear on further stretching will therefore be limited. HEC on the other hand will be in a somewhat more coiled state under low shear resulting in uncoiling under higher shear. MC derivatives are even more coiled and because of the hydrophobic nature of the methyl groups the molecules are stiffer and less easy to uncoil, resulting in a flatter plot (Fig. 7)

Figure 7

Effect of substituent on rheology

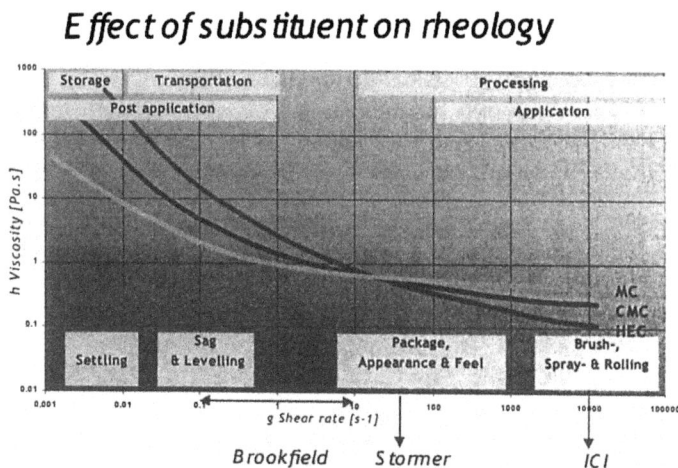

In addition, the molecular weight of the rheology modifier has an important effect on the flow profile. In Fig. 8 two paints are thickened with low and high molecular weight HEC towards the same Stormer viscosity. Because of the increased amount of coiling and the more intense entanglement one can expect a larger pseudoplastic effect for the high molecular weight HEC resulting from its re-orientation and stretching.

Figure 8   *Effect of molecular weight on rheology*

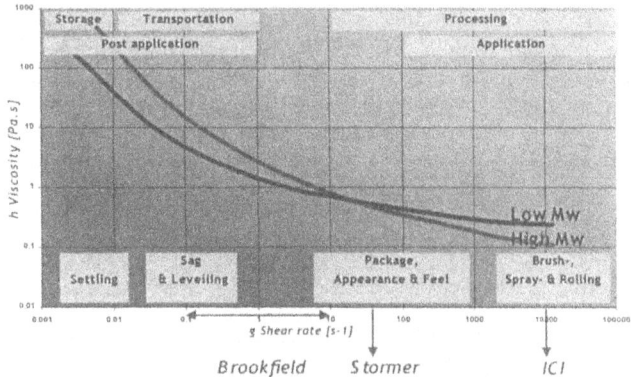

Thickening efficiency, for regular cellulose ethers, is strongly linked to the molecular weight. As can be deduced from what has been presented earlier, it is very important to define the shear rate, which correlates with the given thickening efficiency. Generally the Stormer viscosity of 100 KU is used as the reference.

Since the least pseudoplastic behaviour is associated with the lowest molecular weight, it is easy to understand that Natrosol 250 LR (low Mw) provides the highest ICI viscosity, which makes the paint not easily workable under high shear conditions. However, it does provide good levelling in the coating. Sag resistance is good for all coatings, due to the fact that even the lowest molecular weight grade utilized exhibits sufficient elasticity to prevent the coating from sagging.

6 REDUCED SPATTER CELLULOSIC THICKENERS

It is this same elasticity that causes the poor spatter behaviour of the coatings thickened with the higher molecular weight material. Spatter is caused by elasticity, and elasticity in its turn is strongly related to the molecular weight of the thickener (Fig 9). The higher the molecular weight, the more spatter. Unfortunately the higher molecular weight thickeners are at the same time also the most efficient ones, so the challenge is to have an efficient thickener providing good spatter performance once formulated into paint.

Figure 9

The Origin of Roller Spatter

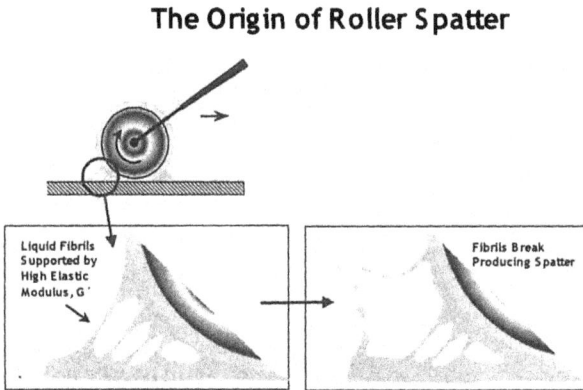

To meet this challenge, hydrophobically modified hydroxyethyl celluloses, abbreviated as HMHEC, were developed and launched. These materials have an architecture similar to HEC, being extended by the addition of hydrophobic groups. By making use of other viscosity increasing mechanisms than just by thickening the water phase, the molecular weight could be decreased, and thus the elasticity, without affecting the thickening efficiency."Regular" cellulose ethers increase the viscosity of the water phase by the formation of hydrogen bonding with water molecules and by means of coil overlap and entanglement.However, HMHEC utilizes additional mechanisms. In addition to the formation of hydrogen bonding with water molecules, coil overlap and entanglement the material is also able to increase viscosity by network building via inter- and intra-molecular bonds, which makes it a versatile polymer (Fig. 10).

Figure 10 *Water phase thickening...HEC vs HMHEC*

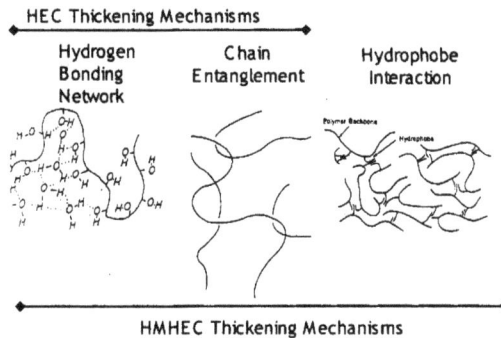

HMHEC is also able to associate via its hydrophobic side chains with major components in the paints that are also hydrophobic in nature, such as latex particles. It thus is able to form a network structure, thereby enhancing its contribution to the increased viscosity of the paint (Fig. 11).

Figure 11 *Hydrophobic Interactions in HMHEC*

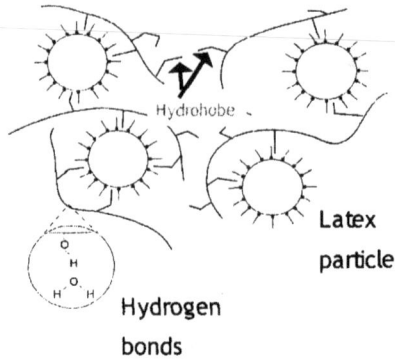

Lowering the molecular weight is not only an effective remedy to control the level of elasticity, but it also changes the flow profile towards a less pseudoplastic character, as has been explained earlier. This in its turn has also a positive effect on other paint properties. In addition to an outstanding spatter resistance, the material provides higher brushing viscosity but also better levelling when compared with higher molecular weight HEC, at equal thickening efficiency.

7 CONCLUSION

The usual cellulosics can be characterised as being the workhorses of the industry when it comes to flat and semi-gloss waterborne paint formulations. (E)HEC, CMC and MC types provide their specific rheology profiles while the associative (E)HEC versions provide good enhanced formulations. Synthetic rheology modifiers like HEUR's and HMPE's rely fully on their associative behaviour. These rheology modifiers can function because of the characteristics of other paint ingredients and their efficiency is thus very much formulation dependent. Blending them with (associative) cellulosics is a proven means to make formulation more robust and enabling them to be more tolerant to changes in raw materials. These factors combine to make cellulose ethers key elements for complete rheology solutions.

WATER BASED DISPERSANTS – THE HIGH PERFORMANCE CHALLENGE

Ian D Maxwell
Avecia Ltd, PO Box 42, Hexagon House, Blackley, Manchester. M9 8ZS

1 INTRODUCTION

Water-based coatings are now well established in Europe in most paint industry sectors, including the demanding automotive market. The same trend is apparent in the USA and more recently in Asia, albeit with a slower rate of change. Meanwhile, the performance requirements are increasingly more demanding, particularly in the areas of dispersion of difficult pigments, better water resistance and more robust formulations.

Polymeric dispersants have been available for some years with improved performance over traditional surfactants. Surfactants, in aqueous systems, typically have a two-part structure – a hydrophilic (water-loving) chain for steric stability and a hydrophobic group for pigment wetting or anchoring. Steric stability is achieved by providing a polymeric layer around the pigment particles which overcomes the normal tendency of the particles to associate with each other. In water-based systems electrostatic stabilisation is also often used since a charged double layer will provide inter-particle stability, particularly in the wet paint. The surface activity of these surfactant groups leads to a range of behaviours (foaming/defoaming, wetting/dispersing, emulsification and solubilization) of which some are desirable and some detrimental to coating performance.[1] The continuing developments in high molecular weight polymeric dispersants are aimed at improving and maintaining the desirable characteristics whilst minimising or eliminating the negative aspects, particularly with regard to water sensitivity. This paper will discuss recent developments in water-based additives from a perspective of improving problem formulations eg. higher jetness blacks, improved rheology and transparency of transparent iron oxides and improved aluminium flake alignment. Dispersion stability and water resistance were also studied to ensure optimum properties from a new series of water-based additives and this will be reported in future publications.

2 OVERVIEW OF WATER-BASED DISPERSANTS

A dispersant is a term referring to specific additives that improve the dispersion of solid particles in a liquid medium. Traditional surfactants encompass this behaviour and a very broad class of materials falls within this general group.

The term surfactant is a contemporary name for surface-active agents, which as the name suggests, are able to modify the properties of an interface that they come into contact with. By lowering the surface or interfacial tension at an interface, eg. solid / liquid, liquid/ liquid or liquid / air, the molecules can bring about a degree of compatibility, and hence stability between the two different phases. This surface activity arises because the molecular structure of surfactants encompasses two groups of very different solubility or opposite polarity.

For dispersants, the key properties are the wetting, dispersing and stabilization of solid particles, such as pigments, into the liquid. After the initial wetting of the pigment surface, the particle agglomerates formed by pigment production processes, are further broken down by mechanical milling. The dispersant absorbs onto the freshly created surface forming a stabilising polymer layer. This steric stabilisation helps to create an energy barrier around each particle and prevents the natural tendency of the particles to re-agglomerate due to the attractive London-van der Waals forces. An alternative to steric barrier, especially in water-based systems, is charge or electrostatic stabilisation. The steric and electrostatic stabilisation (DVLO theory) mechanisms have been extensively reviewed[2-5] and a more detailed examination of these mechanisms can be found in these texts.

Many surfactants generate foam and this has benefits in many industrial applications such as fire fighting foams. However, in coatings, foam is a serious disadvantage since it can lead to defects in the final paint or ink film, such as fish eyes or craters. Foam can also lead to poor millbase rheology and reduce the effectiveness of the milling processes. Agitation by mixing or milling is usually the source of included air at the surface of the liquid. In addition, entrapped air within pigment particles can be retained within the liquid millbase. Anionic surfactants seem to stabilise foams better than the non-ionics and also produce more foam[6].

Surfactants are often used by pigment suppliers as treatments to improve the initial wetting of pigments for aqueous systems. Similarly, resin suppliers will use surfactants in emulsion polymerisation processes to improve the colloidal stability of resin particles dispersed in the aqueous phase. In a complex formulation therefore, there may be a number of different surface active species involved which lead to the problems noted before. Most problems can be categorised by surface tension defects (cratering, fish-eyes, crawling, orange peel, Bernad cells) or pigment dispersion problems (flooding, floating, poor gloss or colour development). All of those problems are described in great detail elsewhere[7].

The development of high molecular weight polymeric dispersants has been driven by the need to improve the desirable characteristics of this additive group whilst minimising or eliminating the negative aspects. For example, stronger pigment wetting effects are found with additives containing multiple anchor groups. These are much less likely to desorb from the surface of the dispersed pigment and cause interface problems such as foam.

As water-based paints become much more widespread in use, then the challenges to achieving a direct match to solvent based paints become more acute. For example, automotive coloured base coats in Europe are mainly water based whilst repair paints are still mainly solvent based. A key target for water-based auto paints is to achieve the same colour properties as solvent-based and in a number of areas this still represents a challenge to the industry. This paper discusses three areas where the continuing development of polymeric dispersants can help to meet the high performance challenge.

3 PROBLEM FORMULATIONS

3.1 Black Jetness

Carbon blacks are manufactured by three production processes, furnace black, lamp black and gas black.[2] Each process gives a range of surfaces with gas blacks having acidic groups while furnace blacks are weakly basic.

The smaller particle size blacks have greater jetness (ie less light reflection). The hue of the paint is usually related to the size of agglomerates with the smaller size yielding a desirable blue undertone.

A consistent theme from paint manufacturers is to produce water-based blacks with equal jetness, and hue, to solvent-based paints. The highest jetness blacks have extremely high surface areas (300-550 m^2/g) and are extremely difficult to stabilize effectively. A series of high surface area blacks with different surface chemistries has been evaluated with a range of polymeric dispersants to study optimum jetness.

3.2 Dispersion of Transparent Iron Oxides

Another small particle size pigment with consequently high surface area, which is difficult to stabilize in water-based media, is transparent iron oxide. High pigment loadings are difficult to achieve in practice with viscosity stability on storage a further issue. Formulators tend to compromise on pigment loading to achieve stable viscosity profiles but this approach will limit transparency and brilliance.

3.3 Aluminium Flake Alignment

Aluminium flake alignment is optimum when the particles are aligned parallel with the substrate to give highest reflectivity and travel from bright, at 90 degrees, to dark as the viewing angle is altered, the so-called flip-flop effect. For water-based paints good flake alignment is achieved at low resin solids.

Clearly it is desirable to achieve good flake alignment at the highest possible formulation solids to minimise drying time.

4 TEST METHODS

4.1 Zeta Potential

Dispersion stability can be quantified by measurement of the zeta potential. The parameter of zeta potential is a measure of the magnitude of the repulsive or attractive forces between particles. Since most inorganic particles, and organic particles by additive absorption, will possess a surface charge in water, a diffuse layer of opposite charge is set up which acts as a single entity around the pigment particle. The plane at the boundary is known as the surface of hydrodynamic shear (or slipping plane) and the potential at this boundary is known as zeta potential (Figure 1).

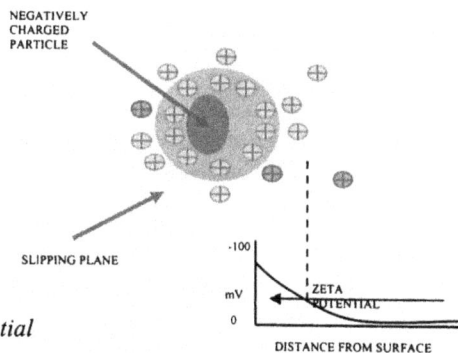

Figure 1   *Zeta potential*

The magnitude of the zeta potential is affected by the degree of the surface charge and/or by the nature of the stabilizing dispersant. However, this only provides information on the aqueous phase and often steric stability is more important in the drying phase as water is removed. Compatibility with the binder system becomes the key to improved properties in the final coating.

4.2 Black Jetness

Jetness has been measured by the Degussa jetness method [8] derived from the following equations in Fig 2:-

$$Jetness: My = 100 \log (100/y)$$

$$Hue \quad : dM = 100 (\log Xn/X - \log Zn/z)$$

$$dM > 0 \text{ (blue)}$$
$$dM < 0 \text{ (brown)}$$

Figure 2 *Jetness (My) and Hue (dM) formulas*

The My and hue values are derived from the CIELAB co-ordinate Yn, Xn, Zn and allows a numerical quantification of black jetness. This is only possible with the development of colour measurement equipment that can operate under reflection conditions of less than 1%. (Figure 3).

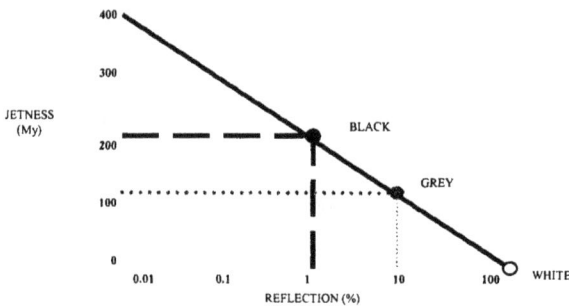

Figure 3 *Jetness variation with % reflection*

4.3 Aluminium Flake Orientation

Flake alignment can be assessed by colouristic measurements at different angles of incident light to the observation angle. The lightness value, L*, is measured as a function of the gloss angle difference (20^0 to 70^0).

The lightness flop or flop index can be defined as flop index $= 2.69 (L*_{20} - L*_{70})^{1.11} / L*_{45})^{0.86}$.

A high flop index number means a better light dark flop effect which is colouristically desirable. [9]

5 BLACK JETNESS RESULTS

5.1 Dispersant Anchor Evaluation

A range of dispersants shown in Table 1 were evaluated on the black pigments also shown in Table 2. Dispersant E is included here for completeness but is evaluated in the iron oxide system.

Table 1

Avecia Dispersant	Anchor Group	Stabilisation Mechanism
A	Hydrophobic copolymer I	Mainly electrostatic
B	Basic	Steric
C	Hydrophobic	Steric
D	Hydrophobic copolymer II	Mainly steric
E	Acidic	Steric

Table 2 *Published data on high surface area pigments*

Pigment	pH	Surface Area (m^2/g)	% Volatile	Type
FW 285	4	350	5.0	Gas Black
FW200	2.3	550	20.0	Oxidised Gas Black
Raven 5000 U2	2.5	583	10.5	Oxidised Furnace Black

All dispersants were tested at 70% active dosage on weight pigment in a resin-free millbase formulation, milled for 2 hours in a Skandex mill. Results of pigment loading, viscosity, gloss and jetness are summarised in Table 3. Gloss and jetness were evaluated in a Bayhydrol 2952 / Cymel 325 stoving water based PU/MF coating.

Table 3 *Dispersant evaluation on high surface area blacks*

(a) FW 285 (70% Aowp) *	A	B	C	D
Pigment Loading (%)	17.5	17.5	17.5	17.5
Viscosity (Pa.s)32s^{-1} (low shear)	3.63	**	0.36	1.30
2392s^{-1} (high shear)	0.27	**	0.06	0.36
Gloss Properties 20^0	42.4	-	50.1	59.3
60^0	74.5	-	79.5	84.3
Haze	337.0	-	296.0	227.0
Jetness My	246.5	-	272.3	285.1
dM	-6.2	-	10.2	14.6
Mc	240.3	-	282.5	299.7

(b) FW 200 (70% Aowp)		A	B	C	D
Pigment Loading (%)		22.5	22.5	22.5	22.5
Viscosity (Pa.s)32s^{-1} (low shear)		1.82	22	1.34	0.06
2392s^{-1} (high shear)		0.141	*	0.11	0.06
Gloss Properties	20^0	57.3	-	55.6	57.4
	60^0	86.2	-	87.5	86.5
Haze		245	-	298	289
Jetness	My	273.0	-	270.3	288.9
	dM	3.52	-	1.98	14.9
	Mc	276.5	-	272.3	303.9

(c) Raven 5000U2 (70%Aowp)		A	B	C	D
Pigment Loading (%)		27	27	27	27
Viscosity (Pa.s)32s^{-1} (low shear)		29.4	34.2	1.90	0.30
2392s^{-1} (high shear)		0.51	0.10	0.21	0.14
Gloss Properties	20^0	-	-	50.9	59.5
	60^0	-	-	82.8	86.5
Haze		-	-	322	305
Jetness	My	-	-	282.7	302.7
	dM	-	-	2.91	14.6
	Mc	-	-	285.5	317.3

*   Aowp = additive on weight pigment
** viscosity too high for milling

The optimum target properties on high surface area blacks are high pigment loading at a flowable viscosity, stable dispersions and good film properties (glass, haze) combined with a high jet black colour.

Results for the three pigments tested show some differences with respect to pigment loading and millbase viscosity. Clearly Dispersant B shows poorer performance in this formulation and the pigment loading would have to be reduced to achieve a similar result to A, C and D dispersants. Dispersant A also shows mixed performance, especially on Raven 5000U2 whilst Dispersants C and D are better matched.

The fact that pigment loading is different between the three pigments is linked to differences in pigment surface chemistry (Table 2) and the differing effectiveness of the anchor group to interact with the surface and stabilise those small particles. Improved jetness is shown by a high value for My, improved blue undertone by high dM and overall jetness by the Mc result.

Clearly Dispersant D gives the best viscosity and gloss properties and in particular improved jetness across the three pigments tested. This suggests that a complex hydrophobic copolymer anchor group can be designed to cover different types of black surface chemistry, rather than the specific acid-base interaction that has been a common approach in the past.

As a further check on comparative performance, dispersant D was compared to three competitive dispersants offered for black pigments in water base applications.

5.2 Dispersant Evaluation

The dispersants evaluated are shown in Table 4.

Table 4 *Competitive dispersant evaluation*

Avecia Dispersant D
Competitor (A) Dispersant 1
Competitor (B) Dispersant 2
Competitor (C) Dispersant 3

Again the dispersants were tested on the three pigments at 70% active dosage on weight pigment in a resin-free formulation. Results are shown in Table 5 for viscosity, zeta potential, gloss, and jetness.

Table 5 *Dispersant Evaluation on high surface area blacks (70% Aowp)*

FW285		*D*	*1*	*2*
Pigment Loading (%)		20.0	20.0	17.5
Viscosity (Pa.s)32s^{-1} (low shear)		0.31	3.6	3.31
2392s^{-1} (high shear)		0.15	3.6	0.40
Zeta Potential		-36.2	-25.8	-26.7
Gloss Properties	20^0	72.4	57.5	61.1
	60^0	85.6	78.0	83.0
Haze		48	147	106
Jetness	My	285.1	267.1	272.8
	dM	14.6	5.2	3.45
	Mc	299.7	272.3	276.2

FW 200	D	1	2	3
Pigment Loading (%)	22.5	22.5	22.5	22.5
Viscosity (Pa.s)$32s^{-1}$ (low shear)	0.06	1.43	3.31	*
$2392s^{-1}$ (high shear)	0.06	0.27	0.34	*
Zeta Potential	-33.8	-24.2	-23.5	-
Gloss Properties 20^{0}	73.9	66.4	61.3	-
60^{0}	86.5	82.4	78.4	-
Haze	59	106	123	-
Jetness My	288.95	281.5	275.6	-
dM	14.95	14.0	11.2	-
Mc	303.9	295.5	286.8	-

Raven 5000 U2	D	1	2	3
Pigment Loading (%)	27.0	27.0	27.0	27.0
Viscosity (Pa.s)$32s^{-1}$ (low shear)	0.30	1.16	28.2	19.8
$2392s^{-1}$ (high shear)	0.14	0.31	0.61	0.6
Zeta Potential	-41.0	-36.8	-	-
Gloss Properties 20^{0}	62.1	37.0	-	-
60^{0}	84.0	69.8	-	-
Haze	187	288	-	-
Jetness My	302.7	279.0	-	-
dM	14.6	4.7		
Mc	317.3	283.7		

* viscosity too high for efficient milling

Only dispersants D and 1 (competitor A) meet the pigment loading / viscosity criteria across the three pigments. Dispersion stability as measured by zeta potential shows less differentiation between the samples with the exception of dispersant D, which shows a consistently higher figure. Generally speaking the higher the absolute value of the zeta potential, the more stable the system will be. A value of 25 mv (positive or negative) is usually required to maintain the dispersed state, particularly so if electrical charge is the sole stabilisation mechanism. In this case the stabilisation is mainly steric in nature.

The jetness properties of dispersant D are also the best across the three chosen pigments, which indicates excellent compatibility of the steric stabilisation chain with the binder system and that particle stability is maintained in the dry film after removal of water.

In summary, this study has shown the importance of the selection and design of the anchor group which can allow broad effectiveness across a range of carbon blacks of differing surface chemistries. Zeta potential has also indicated useful predictive data on the likely ranking of dispersants and related dispersion storage stability. The trend to high jet blacks and a closer match between solvent-based and water-based paints should be possible given new developments in dispersant design.

6 TRANSPARENT IRON OXIDES

Previously reported work has shown that acid-based anchor groups give the optimum bonding to inorganic surfaces, particularly if no additional treatments are in place. This study looks at a range of dispersants on untreated grades of transparent iron oxides with the formulation shown below

Table 6   *Formulation for Competitive dispersant evaluation*

	%
Transparent red iron oxide AC 1000	
Dispersant	55
Water	8
Antifoam	27.86
Humectant	0.14
	5.0

Dispersants were evaluated in the formulation shown in Table 6 using Johnson Matthey pigment AC 1000, and the results are shown in Table 7.

Table 7

Dispersant	Anchor	Pigment Loading	Dosage %	Viscosity $(32s^{-1})$	$(2392s^{-1})$
D	hydrophobic	55	4	41.8	0.66
E	acid	55	4	1.75	0.11
E	acid	55	8	0.12	0.06
Competitor (2)					
Dispersant 4	-	55	8	35.5	0.57

The formulations based on dispersant E, which showed the best dispersion rheology, and the leading competitor dispersant 4 were evaluated further with respect to storage properties, where rheology was re-measured after 6 weeks. The results are shown in Figure 4.

The rheology plots show that Dispersant E gives a stable, non-gelling dispersion with no changes in viscosity after 6 weeks storage at 40^{0}C. Dispersant 4, on the other hand, at this pigment concentration, gelled and no viscosity could be measured after 6 weeks. Showing its tendency to gel on storage.

The gloss properties of paints made from these stored dispersions are shown in Table 8. The transparency and high gloss achieved with dispersant E reflect the excellent dispersion properties achieved on this difficult pigment. Again, careful selection of the anchor group and the stabilising chain can lead to dramatic improvements in dispersion performance.

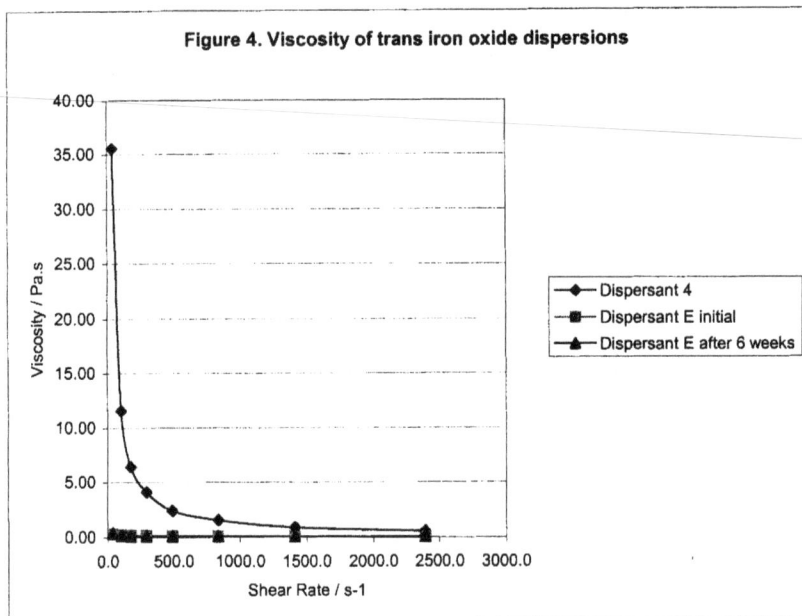

Figure 4. Viscosity of trans iron oxide dispersions

Table 8

	Dispersant E	*Competitor 2 Dispersant (4)*
Gloss 20^0	88.7	20.4
60^0	97.1	59.8
Haze	47.6	438
Transparency	CONTROL	-4

7 ALUMINIUM FLAKE ORIENTATION

The formulation in Table 9 was used to compare the effect of dispersants on a standard water-based aluminium formulation.

Since most aluminium formulations contain waterbased thickeners to avoid settlement of the large particles, the dispersant/thickeners/resin solids interaction is important in producing the optimum rheology curve whilst achieving good flake alignment at the highest acceptable resin solids.

Rheology results are shown in Figure 5 for combination of dispersants with a commercially available HASE (hydrophobically modified alkali swellable acrylic emulsion) thickener.

Table 9

		Water-based Aluminium Formulation
(a)	Stapa Hydrolan 2156	4.65
	Water	4.65
	Dispersant	2.5
	Surface tension modifier	2.5
(b)	Bayhydrol 2952	27.05
	Water	13.51
	Butyl glycol	1.35
(c)	DMEA (10% in water)	4.5
	Thickener	2.52
	Water	13.7
	Cymel 325	3.37
(d)	Butyl glycol	1.35
	Water	4.45
(e)	Water	<u>13.9</u>
		100

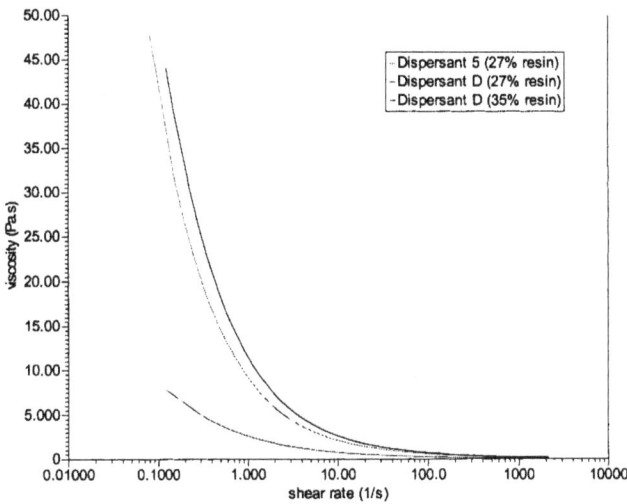

Figure 5. Viscosity of aluminium flake dispersions at different resin solids

The preferred rheology is achieved at 27% resin solids with Dispersant 5 and at 35% resin solids with Dispersant D.

Table 10

	% Resin Solids	20⁰ Gloss	60⁰ Gloss	Flop Index
Competitive Dispersant (5)	27	17.4	47.0	12.4
Dispersant D	27	18.9	50.4	13.1
Competitive Dispersant (5)	35	13.5	42.9	11.6
Dispersant D	35	18.5	50.8	15.2

Table 10 shows the effect of dispersant and resin solids combination on the flake alignment measurement (flop index). Dispersant 5 shows no improvement in Al alignment as resin solids is increased, and this is a fairly typical result seen in many systems. Dispersant D however shows a moderate, but desirable, increase in Flop Index at higher resin solids, which will lead to improved drying time for the final coating.

7 CONCLUSION

Improved additive design can help formulators address a range of difficult water-based formulation issues. The examples shown cover real industry needs as water-based paints replace solvent-based paints in more demanding industrial applications.

A number of application test methods are discussed covering black jetness measurement, aluminium flake alignment and the usefulness of zeta potential is demonstrated in assessing dispersion stability.

8 REFERENCES

1. D Thetford – 'Applications of Oligomeric Surfactants in Coatings' chapter in 'Surfactants in Polymer, Coatings, Inks and Adhesives' – Sheffield Academic Press. To be published.
2. J L Lynn Jr., B. H Borg, Kirk Othmer Enclyclopedia of Chemical Technology, 4ᵗʰ Edn., Vol 23, M. Howe-Grant (Ed.), Wiley, New York, 1997, 478-541.
3. G. Bognolo, Surface Active Behaviour of Performance Surfactants, Annual Surfactants Review, Vol 3, D. Karsa (Ed), Sheffield Academic Press, 2000, 1-65.
4. D. R. Karsa, Spec. Publ, R. Soc. Chem. 1987, 59, 1-23.
5. K. R. Lange, Surfactants, K. R. Lange (Ed), Verlag, Munich, 1999, 144-70.
6. M. J. Schick, E. A. Berger, JOCCA 1963, 40, 66-8.
7. J. Bieleman, in Additives for Coatings, J. Bieleman (Ed), Wiley-VCH, Weinheim, 2000, 65-99.
8. Degussa Technical Information – Pigment Blacks in Waterborne Paint (TI 1227).
9. M Osterhold, ECJ. 2000, 04,406.

ADDITIVES RULES: THE CLASSIFICATION AND SELECTION OF ADDITIVES FOR COATINGS

Jon Graystone

Paint Research Station, 8 Waldegrove Road, Teddington, Middlesex TW11 8LD, UK

Additives play an indispensable role in the performance of commercial coating compositions. A listing of additives by functional terminology shows them to span every aspect of wet and dry coating behaviour. Moreover they are underpinned by many scientific and technical disciplines such as rheology, surface energy, biochemistry, electrochemistry and many others. Formulators must deal with the principles, procedures and 'rules' that govern the selection of additives and the levels at which they are to be used. Formulation with additives is often carried out on an experimental basis, but this is greatly enhanced if the underlying principles are understood. Classification is a first step to such understanding and provides the basis on which rules driven formulation can be based. Such rules can then be used with modern knowledge management tools.

1   INTRODUCTION - ADDITIVES AND FORMULATION

An 'additive' is defined in Standards terminology as "Any substance, added in small quantities to a coating material to improve or modify one or more properties". This says something about what they are not, rather than what they are! They are not the major ingredients yet their role in formulation is vital. Few, if any, coatings would comprise only pigment and binder. As the coatings industry responds to environmental legislation, and the relentless pressure for operational efficiency it is increasingly important that coatings formulators have a broadly based understanding of the technical principles of each component, and the interactions between them.

Formulation may be described as "The science and technology of producing a physical mixture of two or more components, with more than one conflicting measure of product quality"

The role of the formulator is to meet business needs by designing a product that has the required appearance, and a balance of properties suited to the envisaged application. Most coatings must satisfy aesthetic and functional considerations, which range from storage through application to film durability. Making an informed selection of components and then combining them in specific ratios and in the correct sequence at a specified state of subdivision meet these requirements. But the formulator must also contend with the second of the key elements implicit in the formulation definition, namely the certainty that there will be 'conflicting measures of product quality'. This is true of any man made artefact. In a motorcar for example one cannot maximise both acceleration and fuel consumption and a trade off must be made which would be different between a sports car and a family saloon. In coatings there are analogous choices, for example finding a rheology to maximise brush mark levelling without sagging or settlement, and the ever present need to maximise key properties at minimum cost. The number of combinations and permutations that can be made from even a small subset of raw materials is quite staggering, and formulators need to take full advantage of computers in data analysis, optimisation and information retrieval.

Although from time to time it will be necessary to start formulation from a 'green field' situation, by far the most common practical approach to formulating is to modify an existing formulation through substitution of one or more components. Such an approach recognises that an established formula encapsulates a considerable amount of know-how. The process is most effective if there is a-priori knowledge to guide the selection of components and a structured approach to experimental design. Analysis of existing formulae in terms of volume relationships such as the pigment volume concentration (PVC), surface areas, morphology etc. is a means to make the underlying rules more explicit and maximises the learning from past work. Formulation at its best is a professional discipline which requires:-

- Access to all available knowledge including historical formulation and relevant technical data.
- Experience with Experimental Design, Mixture theory and Optimisation routines.
- Conceptual models relating coating structure and morphology to target properties (e.g. pigment packing, composite material science, colloid science).
- In depth knowledge of potential formula components at functional, physical and chemical levels.
- Understanding of physical phenomena such as -
                    - Mechanical Properties, Adhesion
                    - Durability (internal, external, weathering)
                    - Film Formation
                    - Optical Properties (Colour, Opacity, Sheen etc.)

As noted above, formulation in practice is often a process of modifying an existing formula. All major companies producing paints will have large archives of current and past formulations that can provide viable starting points for new products. With proper records such a formulation database can be interrogated to provide general rules. Such rules are empirical but will be strengthened if a physical model is also incorporated. Needless to say, commercial formulations are highly confidential; other sources include patents and publications from raw material suppliers. However these are unlikely to have the refinement that results from the long development time invested in proprietary commercial formulations.

Paint formulating protocols will differ according to the target market but will usually contain the following elements, some of which will operate in parallel:-

[1]   Agree Specification.

A vital step, and often overlooked. Formulation cannot proceed without a clear definition of the target market and any constraints that may be operating including costs. The specification should also address key wet and dry physical properties.

[2]   Understand the 'property weightings'.

This is an extension of the first step. It means putting some value on properties that are in conflict, in order to arrive at the best compromise. Formulation always involves compromises! Customer consultation is vital at this stage.

[3]   Select a starting point.

For a raw material supplier this will usually mean building products around specific technologies or materials. For the Paint Company this could be a green field situation and implies selecting or synthesising the right binder technology. It could mean selecting a water-borne route (as opposed to solvent-borne, high solids, powder or radcure). Thus the formulator must start with an appreciation of the practical implications and limitations of the chosen generic technology. Does the technology have the capability to deliver the right property mix? In practice the choice of technology may also be constrained by operational factors such as the layout and capacity of existing plant.

[4]   Set the ratios of principal components.

This is a fundamental part of generic formulating. Among the more important ratios are that of pigment:binder and volatile:non-volatile (solids). Selection of P:B or PVC ratios will have a major effect on the properties achievable with a given binder, to a large extent it is the PVC ratio that establishes the property mix for a particular pigment/binder combination, however this must be fine tuned to allow for the packing and flocculation effects that are reflected in the CPVC. Although weight or volume ratios provide a good starting point it must not be overlooked that size and area ratios are also important.

[5]   Select 'minor' components

'Minor' by volume, but not in importance, this includes the 'additives', or 'modifiers', which are the subject of this book. In each case the basis of selection and concentration will depend on the technology and should reflect the mechanistic function and is discussed further below.

[6]   Test and re-test.

Some knowledge of repeatability is necessary to make informed decisions. For certain products ascertaining storage stability, or durability, might become rate-determining factors and should be initiated at the earliest possible stage.

[7]   Fine-tune and optimise.

This is where the final compromises must be made in order to maximise the value of desirable properties within operational and financial targets. Greatly aided by good experimental design and the use of statistical data analysis.

[8]   Colours and Tinting

Most paint systems need to be available in a range of colours, for decorative and refinish paints this will be very extensive. Formulating protocols will often extensively develop a single product and then derive colours on a basis of substitution, which will require a rebalancing of all the components. There are important operational differences between colours that are made by a 'co-grind' route, or by blending or tinting. The latter may be in the factory or at the point of sale.

[9]   Scale-up and manufacture.

This too is a major step in formulation. Scale up is based on a premise of similarities but it will not be possible to keep all geometric, kinetic and dynamic ratios constant, hence differences in properties can arise on scale up that will require changes to the process if laboratory and plant made products are to align.

On the basis of the above protocol the starting point for most formulations, having agreed a specification, is to select the principal components and the ratios in which they are to be combined. Mixture diagrams [1] are a useful way of representing formulating options for all or part of the formulation and can subsequently used to display property contour plots.

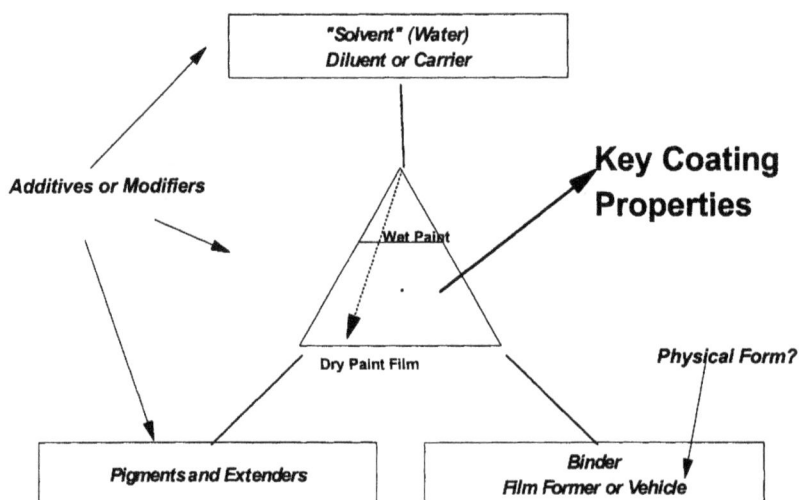

Figure 1 *Paint components and their relationship to Key Coating Properties*

In the case of compositions containing a volatile carrier or solvent, it is enough to specify the volume solids and pigment volume concentration in order to illustrate the essential character of the composition. However, the detail will depend on many other factors such as the physical form of the components. Although there are examples of coatings which are viable using only the main components, by far the majority will require functional modifiers (additives) in order to realise their full potential, or in some cases to be viable. It is at this stage that the formulator must consider the basis on which further additions should be made. A good starting point is to consider functional needs within the context of a particular technical field and supporting chemistry [2] as illustrated in table 1.

Table 1 *Classification of additives by technical, functional and chemical terminology*

Technical Field	Functional Terms	Chemistry (examples only)
Rheology	Structuring Agent, Thickener, anti-settling or anti-sagging, suspension aids, bodying agents. Levelling and Flow Modifiers.	Organoclays (amine mod);Castor Oil Waxes; Metallic soaps; pyrogenic silica; polyamide alkyds; cellulosics, polyacrylates; ethoxylated urethanes; polyacrylamides (associative thickeners). Polyurea.
Electrochemistry	Anticorrosive (in can, in film, flash corrosion).Neutralising and Buffering agents (pH).	Chromates; phosphates; amines; zinc;thiocarboxylic acid salts; basic sulphonates. Amino hydroxy compounds.
Catalysis & Crosslinking	Driers Curing agents	Metal soaps (Cobalt, Zirconium, Calcium) of an octoic acid,dibutyl tin dilaurate; ammonium zirconium carbonate.
Free radical modifiers and complexing	Antiskin. Heat stabilisers, flame retardants. Antioxidants	Oximes and ketoximes. Hindered phenols; ceric and antimony oxide. HALS (tetramethylpiperidine)
Surface chemistry and thermodynamics, Surface Tension	Surfactants. Defoamers, antifoams, anti flood and float, wetting aids, emulsifiers, levelling agents, pigment dispersants. Colorant acceptor Water repellents.	Numerous anionic, non-ionic, cationic and amphoteric agents. Silicones; amphipathic polymers. Polysiloxanes
Solubility parameters	Coalescing Agents Coupling Solvents, freeze thaw stabilisers.	Propylene glycol; benzyl alcohol; glycol ethers; ester alcohols; ethylene glycol.
Electrostatics	Antistatic Electrostatic spraying	Conductive pigments. Polar solvents
Adhesion	Adhesion promoters	Silanes, epoxies
Tribology	Marr and slip aids	Silicone wax. PTFE
Biological and biochemistry	Biocides, bactericides, fungicides, algaecides, antifouling.	Tributyl tin oxide, formaldehye, thiodiazine-thione, oxazolodimes, thiophthalimide.
Organoleptic	Reodorants	Esters.
Optics	Matting agents, spacers, refractive index modifiers.	Silica, blanc fixe, ultrafine TiO_2, polymer beads
Electromagnetic electron transfer	Optical brighteners Photo initiators, synergists UV absorbers	Benzoxazoles. Benzophenone, Thioxanthones Hydroxybenzophenones, tetramethyl piperidine. Benztriazoles. Triazines.

2   ADDITIVE RULES

In order to build up a satisfactory recipe the formulator must apply 'rules' which address such questions as

- What to add?
- How much to add?
- When and how to add it?

The question of 'what to add' is governed in the first instance by functional needs and additives may be short-listed by what are essentially phenomenological rules or 'rules of thumb'. There are many ways in which additives may be further categorised which are discussed in more detail in this book. From a pragmatic perspective, additives may be grouped into three broad categories albeit with some overlap:-

Category:		Examples:
A	Necessary to realise inherent properties	pigment dispersant, catalyst, drier or curing agent, coalescent
B	Confer an additional desirable property	rheology modifier, biocide, adhesion promoter, freeze thaw stabiliser
C	Overcome an undesirable side effect	Anti-foam, anti-flooding, anti-skin

It is clear from this grouping that an all-embracing term like 'additives' can hardly do justice to the many functions required from this grouping. In fact, formulators will describe materials from groups A and B by their functional terms (i.e. pigment dispersant, thickener etc.) rather than the collective term. Arguably it is group C, the 'anti-' and trouble shooting materials that are more likely to be described as additives, and viewed perhaps as a 'necessary evil'. This is because of the problem of 'circular formulating'. A typical example of the latter would be where a wetting agent causes foaming, cured by addition of silicone antifoam in turn causing cissing. Further remedial action then interferes with the pigment dispersant stage, and so on! It is more than anecdotal to say that over a period of time formulations evolve to reflect the input of several formulators and it may be necessary at some stage to remove all the group C additives, and start again.

3   TYPES OF RULE

The logical procedures which lead to the selection of additives can be summarised as:-

Rule Type	Scope
Selection	Based on functional requirements
Concentration	How much to add in order to achieve the desired effect
Environmental	Concerns Constrains arising from legislation
Equilibrium	Deal with kinetic and dynamic aspects
Compatibility	Surface energy interactions and phase solubility, antagonistic interactions in general.
Hierarchical or Priority	Order of addition

Rules of Selection

'Rules of Selection' must be based on good logic underpinned by phenomenological, experimental or theoretical knowledge. As an illustration consider the addition of a coalescing agent to a coating based on an aqueous dispersion (latex).

At the phenomenological level it may be known that Solvent 'A' works well with acrylic dispersions while Solvent 'B' is better for vinyl dispersions. Experimentally the important question of how much to add can be determined by systematic experimentation. However, for fine tuning some theoretical knowledge about the influence of transient solvent on the glass transition temperature (Tg), and the influence this will have on the minimum film forming temperature (MFT) is useful. This would undoubtedly lead to some consideration of the distribution (partitioning) of the solvent between the aqueous and polymer phases, which can be guided by considering the solubility parameter and its hydrogen-bonding component. Other factors to be taken into account include evaporation rates since residual solvent, if retained, can be detrimental to film properties. At this stage there would be more than enough information for practical paint formulation even though there is still considerable scope for in-depth investigations. The work and slightly contrasting views of Chevalier [3] and Denkov [4] on the detailed mechanisms of film formation are illustrative of this, while Feng and Winnik [5] make interesting observations on the kinetics of drying and the consequences of bi-modal and mixed polymer dispersions.

Concentration Rules
For the practical formulator in a development laboratory there is a pressure for quick solutions and the complex nature of commercial formulations, with the possibility of many interactions, does not lend itself to fundamental study. Large companies may therefore carry out applied research using model systems, but on completion the results are translated into a simplified rule such as "when using acrylic copolymers use a blend of coalescent A:B such that the total level is x% expressed on the total volume of polymer". This kind of statement is a good example of a pragmatic rule where, as a result of prior investigation, the concentration of an additive is expressed as a percentage of a component with which it must have an interaction. Another example would be to add drying catalysts to oil modified alkyds as a % weight of metal with respect to the non-volatile autoxidisable component. For other ingredients such as dispersants it might be more appropriate to base the amount on a parameter such as pigment area.

These examples of 'Concentration Rules' are a convenient short cut for formulators but there are clearly dangers, not least that the basis for the rule becomes obscure and exceptions cannot be dealt with. There is also the issue of partition and dilution such that the concentration of the active ingredient in relation to its 'target' is not straightforward.

Priority Rules
There are many examples where the role of an additive defines its entry point into the formulation. A wetting agent, not to be confused with a pigment dispersant, should clearly be present at the early stages of pre-mixing. Heat sensitive materials such as some biocides may need to be added after cooling.

Practical problems also arise from concentration gradients changes during paint manufacture. These might be described as 'Rules of Priority', which have been derived to avoid pigment and other kinds of 'shock'. Daniel's rules (e.g. "grind at low polymer concentrations and let down at high" [6,17]), are rules that have served well and may be explained in terms of pigment/binder/solvent affinities [7], or more fundamentally by thermodynamic second law principles invoking free energy, entropy and enthalpy as discussed under 'compatibility' below.

Environmental Rules
In formulating for today's products it is not enough to identify additives, or other raw materials, solely by their functionality. There is an increasing level of legislation focussed on safety, health and environmental issues. Some indication of the breadth of these issues is given in the following list:-

* VOC (or 'Solvents') Directive
* Directive on Indicative Occupational Exposure Limit Values
* EU Ozone Depletion Substances Regulation
* Transport and Packaging Regulations
* CHIP 3 (Chemicals Hazard Information and Proposals for Supply) Regulations 2002

- Packaging Waste Directive
- Contaminated Land Groundwater Regulations
- Climate Change Levy
- COSHH Regulations (Control of Substances Hazardous to Health)
- Dangerous Preparations Directive
- Biological Products Regulations
- Extension of the validity of ecological criteria for paints and varnish eco-label
- Adaptation to Technical Progress of the new Dangerous Substances Directive
- Amendment to the Safety Data Sheets Directive
- Amendment to Directive on plastics materials and articles to come into contact with food
- Decorative Paint Directive
- Coatings Care/Responsible Care
- Toys Regulations

Such rules act as a constraint on the selection of additives and place a strong requirement for material information on specific additives. This may be provided through material data safety sheets (MSDS) and other means. Fortunately such information is increasingly available through Internet web sources. When analysing a formulation from the perspective of environmental rules it is important that the constituents of additive mixtures, present in many proprietary products, are broken back into their constituent parts. For example some additives contribute to the VOC content of the final product.

Equilibrium rules
Another aspect of paint formulation arises from kinetic effects, and the time that may be taken to reach equilibrium. For example Young's equation is used to quantify aspects of wetting including penetration and spreading. However the equation assumes equilibrium, and the fact that surfactants will take a finite time to migrate to a new surface makes it necessary to consider dynamic as well as static surface tensions in surfactant selection. 'Rules of Equilibrium' are also relevant to situations such as pigment dispersion where low molecular weight materials that are initially adsorbed onto a pigment surface are gradually replaced by slower migrating but more thermodynamically stable high molecular weight materials, particularly if the latter have multiple anchor points. This aspect explains why fresh and aged mill bases may behave differently on let-down. Such differences can be relevant to apparent scale differences between laboratory and production. Equilibrium rules may also be invoked when letting down pigment dispersions with aqueous or non-aqueous polymeric dispersions. Pre-dosing with components of the mill base surfactant system will reduce shock effects and in some cases avoid total flocculation.

Compatibility rules
The issue of 'compatibility' provides a further illustration of both heuristic and theoretical rules that underpin formulation and additive selection. The components of a coating will all have a positive or negative affinity for each other that can be related to intermolecular forces and thermodynamics and broadly grouped under the heading of 'surface energy' [7].

    'Rules of Compatibility' may be invoked to aid selection. Full compatibility between components is often seen as an advantage e.g. in maintaining clarity and full gloss during film formation. However, there are many circumstances in which intermediate or complete incompatibility are advantageous. An example of the latter would be seen in coatings where enforced stratification gives beneficial film properties such as improved priming [8]. Likewise pigment dispersants are often designed to have an intermediate compatibility so that the stabiliser is neither too soluble in the liquid phase nor inactivated as a steric barrier by being too strongly adsorbed onto the pigment. Such reasoning explains another of Daniel's rules "disperse in poor solvents and let down in good" [17].

Compatibility rules can simply be encoded into narrative terms such as 'coupling agent' and illustrated with phase diagrams. Terms like 'polar' and 'non-polar, or 'hydrophobic' and 'hydrophilic' go one step further in qualifying compatibility issues. Solubility parameters [20] have been an invaluable way of quantifying thermodynamic factors through a 'cohesive energy density function' to provide a basis for selecting solvents as well as characterising polymers and pigments. The HLB scale is a good example of a rules base where the semi-empirical protocol of Griffin [9], has been extended with a more theoretical basis using group contribution theories [10]. Other concepts useful in the selection of dispersants for pigments are acid-base [18] and donor-acceptor models [19].

In the case of water-borne coatings the charge on the pigment surface will have a major effect on stability, either by direct charge stabilisation, or by assisting the anchoring of steric stabilisers. Considering the pH relative to the isoelectric point of the pigment can derive simple but important rules. If the pH is below the isoelectric point the charge on the pigment will be positive, and vice versa. pH will also control the degree of dissociation of the dispersant and hence the propensity for adsorption.

4   KNOWLEDGE BASED SYSTEMS

On the basis of the above discussion it is clear that paint formulation in general, and the use of additives in particular is based upon the application of various rules which vary from the descriptive and empirical, to those which have a sound theoretical basis. For a company involved in formulation, the effective deployment of appropriate rules would constitute part of the skills base underpinning core competences [11]. It is therefore important for companies that this knowledge is well managed to nurture both individual and organisational learning. Learning has been defined as increasing the capacity to take effective action [12] and there are now available a broad range of tools which can aid this process. Which of these is the most appropriate will depend on the nature of the organisation and the task in hand. Tools available to aid the process of formulation can be classified under such headings as, databases, expert systems, statistical techniques, neural networks, algorithms and models. They are not mutually exclusive.

Databases:
All organisations use databases to store operational data including formulations. For many years now these databases have been computerised and in many areas the 'relational database' has superseded the traditional hierarchical database. The former offers many tools for interrogation but it is often useful to download information and use high level programming aids to transfer and manipulation data from different sources. The ubiquitous spreadsheet is a powerful tool for formula analysis and can also be used for formula synthesis. Spreadsheets can readily incorporate the 'rules of concentration', which go to building up a practical formulation [13], and can incorporate calibration data derived from systematic experimentation. Current releases of spreadsheets have powerful mathematical functions that can enable the compositions meeting defined property criteria to be calculated.

Full text databases are available from many sources, and will be supplemented by internal sources. With the increased amount of information available via the Internet there is the danger of a surfeit of information. This kind of knowledge base is divergent which does not always lend itself to problem solving. However, developments in the use of 'hypertext' and content sensitive mark-up languages can provide links between databases and give a conceptual framework that aids understanding. A major European multi-client project 'KnowCoat' [21], is currently designing a new knowledge management platform for the coatings industry, and includes applications specifically targeted to formulation and additive selection.

The solution of formulating problems through additives and modifiers is efficiently carried out with a convergent approach, which can be delivered through the means of an expert system.

Expert Systems:
Expert systems are collations of knowledge and experience used to make consistent decisions in a specific area of expertise. Although usually implemented as computerised programs, simple ones can be represented in text form. Expert systems are not appropriate if a complete algorithmic solution to the problem exists, or if common sense will suffice. They do lend themselves to the applications of a complex rules base and as such can be very appropriate to the selection and use of additives. There are available commercially expert system building programs (shells) which build up knowledge bases on a hierarchy of IF-THEN rules which have been applied to areas such as pigment dispersant selection [14]. Subsequently specialist companies have developed formulating software integrating database and formulation rules that can be adapted to all types of formulation including paint [15]. Such systems have been successfully commercialised in the pharmaceuticals field and elsewhere [20].

Case Based Reasoning:
Whereas the example given above attempts to model an expert's knowledge as a purely deductive process, it has also been recognised that this approach does not fully emulate the way may experts tackle a problem. Rather than going to first principles experts will recognise situations and remember solutions to previous similar problems. Case Based Reasoning [16] attempts to model this problem solving approach. CBR uses algorithms that are not unlike advanced text retrieval systems and as such both data input and questions can be posed in a narrative form. In essence it is an approach to building knowledge that involves reasoning about a current problem based on the retrieval of similar stored cases. In contrast with formal deductive logic it is experience rather than theory, which is the primary source of knowledge, solutions are re-usable rather than unique to each situation and the objective is the best available solution rather than a guaranteed exact solution. In the context of additive selection CBR can be useful in a problem-solving mode, whereas deductive expert systems are more appropriate for rules of selection and concentration.

5   CONCLUSION

Additives, or Functional Modifiers, are indispensable to commercial paint formulation. Their efficient use requires the application of heuristic rules gained through experience but enhanced by an understanding of fundamental scientific principles. Formulation, optimisation and problem solving require a framework, which connects theory with practice and observed phenomena with underlying cause. Provision of information to this end by suppliers, and the use of internal or proprietary knowledge by producers, is greatly facilitated by modern knowledge management techniques and the use of various types of expert system. Although considerable progress has been made in the use of such systems there are still opportunities for exploitation. A new IST initiative on knowledge management offers the coating industry the opportunity to evaluate some of the latest developments.

6   REFERENCES

[1]   Cutrone L."Principles of Paint Formulation", Woodbridge (Ed). Blackie 1991. Ch 5.
[2]   Graystone J A. "Premium on Information as additives become increasingly complex.",
       The Coatings Agenda 1996 (p163-164)
[3]   Chevalier Y et al. Coll.Polym.Sci., 270, 806 1992
[4]   Denkov et al. Langmuir, 8, 31383 (1992)
[5]   Feng J and Winnik M A   Journal of Coatings Technology., 68 (852), 39 1996
[6]   Daniel F K. "Getting the most out of a high speed disperser" . American Paint Journal
       1987, Vol 66 No 46 pp 60 -65
[7]   Schroder J. "Considerations on the energetics of pigment dispersions containing solvent
       and binders". Prog. Org. Coatings, 12 (1984) 339 - 367.
[8]   Carr   C.   "Self   Stratifying   Programmes",   PRA   Research   Project   G9R1
       http://www.pra.org.uk/
[9]   Griffin W C. J Soc. Cosmet.Chem. 1949 1 311
[10]   Davies J T, Rideal E K "Interfacial Phenomena", 2nd Edition, Academic Press, London
       1963
[11]   Graystone J A. "Integrating Colour Delivery Skills". PRA Conference, Leeds University
       3-4 May 2000.
[12]   Kim D H. "Do Organisations Learn?", Chemtech, September 1994 pp 14-21
[13]   Kavanagh P E, Thiedeman W A P. "Paint and Resin Formulation with a Spreadsheet",
       Surface Coatings International 2000 (5) 246 - 249.
[14]   Nelson R. "An expert system for selecting dispersants for industrial use", 7th IACIS
       World Congress, Compiegne France - July 1991
[15]   Logica Formulogic (http://www.logica.com/de/sector/industry/products/index.asp)
[16]   Case Based Reasoning Packages - see (http://www.cbr-web.org).
[17]   Daniel F K, "Flocculation and Related Pigment Interactions in Paints" Journal of Paint
       Technology, 38, No.497, 309 (1966).
[18]   Javier L and Schreiber H P, "Specific interactions and adsorption of film-forming
       polymers", J. Coatings Tech. 1991, Vol 63 No 801, 81-90
[19]   Guttmann V, "The donor-acceptor approach to molecular interactions", Plenum Press,
       1978.
[20]   Colbourn E, Rowe R, "Expert Systems", Chemistry in Britain, July 2001, 54 – 57.
[21]   Knowcoat: Neural Knowledg management solutions for the coating market value
       chain. IST-2001-33262, (http://www.knowcoat.net/).

*Author's Note: This paper is a revised version of one prepared for 'Addcoat 2001' January 2001, Orlando,
USA.*

PRACTICAL ASPECTS OF FORMULATING WITH ASSOCIATIVE THICKENERS

Isabelle Mussard

Rohm and Haas European Laboratories, BP 249, 06905 Sophia Antipolis Cedex, France

1 INTRODUCTION

Nowadays there is a need for cost reduction in the paint industry. The challenge is to maintain margins and reduced raw materials and operating costs at the same time keeping the same quality [1]. Paint manufacturers do not have many parameters to play with to reduce cost and increase plant capacity. One possibility is to find a way to replace dry powders in the formulation, thus the strong willingness to replace cellulosic thickeners. For a long time the paint industry has been trying to replace cellulosic thickeners but until now with the proposed technologies, no product could combine cellulosic-like rheology and the benefits linked to acrylic associative thickeners like low cost, efficiency, quality and liquid form [2]. Rohm and Haas has recently developed a new product line of associative thickeners which represent the first alternative to cellulosic thickeners. These rheology modifiers are designed polymers that provide manufacturers with an appealing new formulating option in this area [2].

2 THICKENERS AND THICKENING MECHANISMS

Coatings formulations based on water borne polymers usually depend on specific ingredients for rheological profile adjustment and overall properties optimization. There is a large variety of additives which can be used to modify the rheology of a system. Natural organic derivatives (e.g. cellulose ethers) have been traditionally used in the paint industry and are well established. Synthetic thickeners, on the other hand, have appeared later but quickly became very popular, mostly because of their ease of use (liquid versus solid) [4].

2.1 Conventional thickeners

In a coatings context, the more common conventional thickeners used for many years are cellulose ethers. Their molecular weights vary from low 50,000 g/mol to very high 800,000 g/mol and directly relate to their thickening efficiency (i.e. low molecular weights thicken less than high molecular weights) [5].

The thickening mechanism of conventional thickeners is the following: the hydrophilic polymer backbone tends to associate with the surrounding water molecules through "hydrogen bonding", increasing significantly the hydrodynamic volume of the polymer itself. There is then less available free space for the particles (latex, pigment, extender) to move freely and, as a result, the viscosity increases.

2.2 Associative thickeners

Rheology modifiers or associative thickeners are hydrophobically modified water soluble polymers. They act both by thickening the water phase and by association with paint ingredients. The relative balance of these two mechanisms depends on the type of rheology modifier. The hydrophobic groups mainly associate with emulsion particles through surface adsorption. This network formation generates a major increase in viscosity, it reduces the mobility of the particles. It also gives a uniform stable dispersion of both pigment and emulsion particles, thus contributing to greater ultimate gloss potential [4].
Their thickening efficiency will depend on the type and the number of hydrophobic groups per polymer chain and the molecular weight.

2.3 Conventional and associative thickeners in paint formulations

When adding a cellulosic thickener to a colloidal system (latex polymer or pigment dispersion), one can observe a phase separation through the formation of two distinct domains : a concentration of solids latex and/or pigment particles - similar to a flocculation - and a solution of thickener and water. This phase separation can be explained as follows:
As said earlier, the hydrodynamic volume of cellulose ether backbone is very large and the polymer chain is rigid. Once in a paint system the thickener backbone is too big to position itself between two adjacent particles and thus remains at the outside. Being more hydrophilic than the latex or pigment particles, it attracts by osmotic pressure the water molecules that are between these particles and then contributes to a reduction of the inter-particle distance, and so on, until an equilibrium is reached with the phase separation status (Figure 1). This phenomenon is reversible and thus cannot be seen as true flocculation, although the adverse effects are similar: structure build-up, yield point, poor flow, low gloss and opacity, etc... Unlike the conventional thickeners, the associative thickeners through their association with the latex particles maintain a more uniform distribution of the latex and the other paint ingredients leading to better flow, opacity, gloss...(Figure 2) [5]

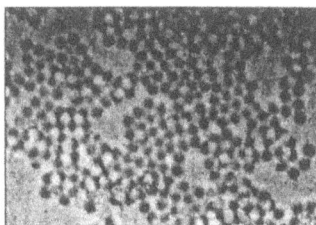

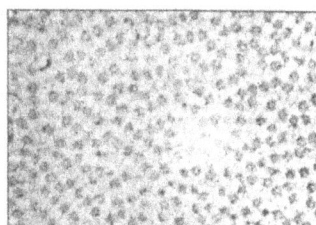

Figure 1 *paint film with non associative thickeners*

Figure 2 *uniform paint film with associative thickener*

3 ADVANTAGES OF THE DEVELOPMENT OF A NEW TECHNOLOGY

Cellulosic thickeners generally are simple to use, tolerant to small formulation changes and they also have good water retention properties [4]. However, they have always had drawbacks as thickening agents. Being supplied in powders, they might be difficult to handle and especially the introduction during the manufacturing process: the plant productivity decreases as the solvation is quite long. Moreover, it is very difficult to use cellulosic thickeners for viscosity adjustment at the end of the production process. They may also generate high formulating cost since their low efficiency requires to use rather large amount of product to reach the desired viscosity. Their sensitivity to microbial attack can alter their thickening efficiency. Last but not least, paints formulated with cellulosic thickeners exhibit poor flow/leveling and are prone to severe roller spattering.

3.1   A new technology

These new rheology modifiers are modified associative acrylic thickeners with a surfactant- like character and an anionic modification which have been especially designed to be a good match for cellulosic thickeners. This technology represents an attractive alternative to a wide range of cellulosic thickeners in formulations from matt to semi-gloss paints using only two products (blending strategy). They provide a higher production efficiency as they are liquid and lower formulating costs due to lower prices and higher efficiency versus cellulosic thickeners. They are also biostable given their polymeric chemistry and used in paints they show excellent application performance and good overall properties [1].

3.2   Blending strategy

The goal is to offer to paint formulators an easy way to convert from cellulosic thickeners to designed rheology modifiers maintaining rheology profile and overall properties.
Thanks to a blending strategy, the two new products offer very interesting solutions. The formulator will select the low molecular weight new product to match rheology profile of a low molecular weight cellulosic. In can appearance and application properties given by high molecular weight cellulosic thickener will be easily achieved by selecting the high molecular weight new product. A proper blending of the two new products will simply replace cellulosics of intermediate molecular weight.
Only two products will be needed to adjust the rheology of a wide range of paint formulations.

3.3   Reduction of formulated and operating costs

The cellulosic thickeners market has had low growth in the past recent years when the demand for liquid polymer type thickeners has been greater. This because of the trend to full liquid production line and easily - tailored systems. Paints manufacturers wait for cost effectiveness from suppliers with improved, more efficient and competitively priced products.
This new technology offers an opportunity for major cost savings. These rheology modifiers are cheaper and more efficient than cellulosic thickeners. Their usage level is 10 to 20 % lower enabling paint manufacturers to decrease raw materials costs [1].
The gain in productivity is also an important part of savings. These rheology modifiers are supplied at low viscosity and are easy to pump especially for bulk handling and automatic metering equipment. Generally speaking, customers notice a 5 to 30 % cycle time reduction of the manufacturing process, in most cases this is due to the elimination of the solvation step required by cellulosic thickeners. This obviously also means an increase of production capacity with no capital investment. The trend is to increase capacity and productivity going through product rationalization but also through the use of all liquid technology to cut down cycle time. This explains why liquid thickeners are taking market

share and why pigment/extender slurries are used more and more. Productivity increase is the main driver for cellulosic replacement.

A practical example from a customer is the switch from a high molecular weight cellulosic thickener in a high PVC paint to its alternative in the new technology. First a 50 % thickener amount reduction brought a 5 % formulation cost savings. During the production, the cellulosic thickener was pre-dissolved in a separate tank. Switching to an associative liquid thickener alternative permitted to reduce the total batch time by 30 %, because this technology does not require premixing before addition. Finally an additional 30 % savings were achieved because of the smaller amount of energy required throughout the whole manufacturing process.

In other types of production with no cellulosic premix, blending cycle times are nevertheless reduced: dry cellulosic thickener has to be added slowly and needs time to swell and build up viscosity; whereas the new product can be added more quickly and reacts immediately.

3.4 Micro-organisms resistance

Most of the time, cellulosic thickeners are pre-dissolved in a separate tank and used under a liquid form in the manufacturing process. In general, this premix is a blend of cellulosic thickener, water, biocide and defoamer. This colloid "solution" is made to be pumpable and large enough to be used in several paint batches.

The main causes of viscosity loss of cellulose ethers are of microbial (bacteria, fungi and enzyme) and chemical (redox process) natures. Biocides are used to kill bacteria and fungi but they are usually not effective against enzymes. The enzymatic liquefaction is known to be the strongest process. Studies have been made on different cellulosic thickeners with bacteria, fungi and commercial cellulase enzyme and the overall conclusion was that cellulase is the most significant cause of viscosity loss [1].

Generally contamination in a colloid "solution" can appear after only 24 hours resulting in a decreased efficiency of cellulosic thickeners and can contaminate the paints which then show visual deterioration like phase separation and the development of a characteristic odour. This issue is a problem for the paint industry as it first leads to an increase in formulating costs linked to larger amount of thickener in order to obtain the required viscosity and secondly to the excessive costs associated with claims when contaminated paints are returned.

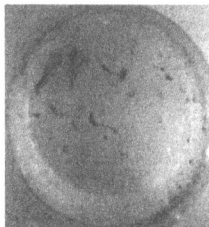

Figure 3 *Colloid solution sample from the surface - visual aspect (x 1.5)*

4 FORMULATING CONSIDERATIONS

Although this new rheology modifiers technology is designed to match the performance of cellulosic thickeners, its acrylic associative character requires a different formulating approach. Due to their thickening effect, these associative thickeners will tend to interact with a variety of components that are present in a paint formulation [6]. The possible influences on rheology modifier interactions include pH, dispersant type and level, composition of the emulsion (hydrophobic or hydrophilic), polymer particle size (degree of association is dependent on total surface area), stabilization of emulsion (colloid or surfactant).

4.1 pH control

These new rheology modifiers have two thickening mechanisms: they interact with the binder and they thicken the aqueous phase due to interactions of carboxylate functions with the molecules of water. These products are not soluble at low pH. Raising the pH with a hard base, they become soluble and they establish the associations, which will build viscosity. This thickening effect starts at pH 6 and the rheology modifier is fully solubilized at pH 7 (Figure 4).

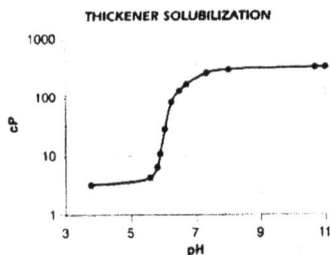

THICKENER SOLUBILIZATION

Figure 4 *Viscosity development of new rheology modifier technology versus pH*

For optimum results, the formulator has to adjust paint pH above 8 before adding this rheology modifier. If the pH is below 8 during the manufacturing process, the thickener can be destabilized and it can lead to grit formation. To ensure stability, it is recommended to add enough base to have a final pH of 8.5 to 9 (any types of bases can be used: ammonia, sodium or potassium hydroxide and amine alcohol derivatives) [6].

4.2 Dispersant

The choice and level of dispersant are critical. The issue with these new products is their ability to interface with other ingredients the same way as dispersants would [6].
Paints containing cellulose might be understabilized as cellulosic thickeners do not interact with other paint ingredients. These new products have a chemical composition similar to dispersants (polycarboxylate functionality) and hence they tend to compete with the dispersant to get adsorbed onto the pigment surface, thus resulting in bridging flocculation, which leads to viscosity instability, levering etc. An appropriate level of polyacid dispersant is recommended (0.5 to 1% depending on total paint PVC) in order to saturate the pigment surface and minimize the competition with rheology modifiers. Hydrophobic copolymer type dispersants will have a tendency to compete with these associative

thickeners and may cause colorant compatibility problems, viscosity instability and syneresis. In high PVC paints, the addition of a wetting agent in the grinding phase will give good dispersion and stability.

4.3 Binder composition

Polyvinyl acetate or vinyl acrylic based paints tend to show more stability sensitivity with this new type of technology. Because of potential hydrolysis of vinyl acetate, the pH could drop below 7.5 and the rheology modifiers would lose their thickening efficiency. Around pH 7 there is a gelation/levering effect probably due to competition for available base between the associative thickener and the other paint ingredients.
In the case of PVA paint hydrolysis, heat ageing stability will need to be carefully checked. The use of alkaline extenders such as calcium carbonates as buffers will definitely improve stability [1].

4.4 Order of addition

This new technology offers the paint manufacturer both ease of handling and flexibility in the thickener use order and method of addition. These rheology modifiers can be partly added to the grind to get optimum pigment dispersion through viscosity control of the pigment slurry and then in the letdown when finishing the paint manufacturing. They can act as a post-additive rheology adjustment without affecting performance properties as long as sufficient neutralizing base and mixing are available. If the mixing conditions are not optimum or if the binder is very reactive (high hydrophobicity), it is recommended to add the rheology modifier more slowly or to pre-dilute it at 50% with water before addition.

4.5 Scrub resistance

Best scrub resistance results could be achieved with the variation of different parameters: an increase of pH level (up to 9), dispersant level (from 0.6 to 1 %) or the introduction of organic opacifier could have a very positive effect on scrub level keeping the same overall properties [8].

4.6 Paint stability

For optimum stability, it is advisable to follow the pH recommendations given above and to adjust the level of high HLB surfactants in the formulation[8].

5 CONCLUSIONS

The introduction of these new rheology modifiers provides paints manufacturers with a very attractive alternative to cellulosic thickeners. These rheology modifiers are the first acrylic associative thickeners that offer a similar rheology, in-can feeling, appearance, and application properties as cellulosic thickeners. They also show some key advantages compared to cellulosic thickeners. Supplied as liquids, they are easier to handle and because of their polymeric chemistry, they are less sensitive to microbial attack [1]. In application trials, professional painters found that new rheology modifier based paint performs similarly to cellulose thickened paint but with a better spattering resistance.

However, what makes this new technology very attractive is that it offers opportunity for major cost savings. These rheology modifiers function more efficiently than cellulosic thickeners giving the possibility to decrease raw material costs. In addition, productivity improvements are possible and have been confirmed by customers.

6   REFERENCES

1   Acrysol DR Technical Data sheet, Rohm and Haas Company, 2000
2   Rohm and Haas Company, *Brush Strokes, Acrysol DR Thickeners : New Tools for Economical Interior Wall Paints*, unpublished data.
3   E. J. Schaller, P. R. Sperry, *Handbook of Coatings Additives*,1992, Vol 2, 105-114.
4   J. Prideaux, *Rheology Modifiers and Thickeners in Aqueous Paints*, , 1-6.
5   B. Lestarquit, *Rheology seminar*, Rohm and Haas Company, unpublished data.
6   Rohm and Haas Company, *Brush Strokes, Acrysol DR : Formulating Considerations*, unplublished data.
7   I. Mussard, *Acrysol DR evaluation*, Rohm and Haas Company, unpublished data.
8   I. Mussard, *Acrysol DR interactions with other paint components*, Rohm and Haas Company, unpublished data.

MATTING AGENTS FOR THE MILLENNIUM

Hans-Dieter Christian

Degussa AG, Advanced Fillers and Pigments, FP-FA-AT2/LA, Postcode: 885-011, Rodenbacher Chaussee 4, 63457 Hanau

1 INTRODUCTION

Synthetic silicas have been used in the paint and coating industries for more than 50 years. Among other applications, they are used to adjust the rheology or to improve the corrosion resistance as well as to matt coatings.

The continuing trend towards matt surfaces remains unchanged. The latest considerations in the automotive industry even foresee formulations of matt topcoats, and the furniture industry continues to favour matt surfaces that bring out the original structure of the wood. Even the traditional markets for glossy furniture surfaces such as those in Italy, Spain and China are no longer a closed shop to the demands of designers.

In order to take further account of this trend, Degussa-Hüls has begun to develop a new generation of matting agents. In this presentation the creation of these new, uncoated products and their application properties are described.

2 MANUFACTURING PROCESSES AND CHARACTERISTICS

The first manufacturing method is the thermal process (Figure 1a). Currently, ACEMATT TS 100 is the only matting agent available in this range. Fumed silicas differ from precipitated silicas and silica gels in three major characteristics:

They have a lower drying loss (determined over 2h at 100°C), which indicates the amount of physically bonded water. This property is especially important when the silica is to be used in polyurethane systems and particularly in moisture-hardening polyurethanes.

They have a lower ignition loss (determined over 2h at 1000°C), which shows the amount of chemically bonded water.

The looser agglomerate structure allows a higher level of transparency to be achieved in most coating systems.

In addition to these three parameters, the average agglomerate particle size – for the sake of simplicity also known as particle size - and also the particle-size distribution are important.

The second method is the wet process, which is used to produce precipitated silicas and silica gels (Figure 1b, 1c). This also allows a wide spectrum of different matting agents to be produced, which in turn enables varying degrees of matting from eggshell to dull matt and textured finishes. In many cases the silicas are subsequently treated with wax to improve their suspension properties.

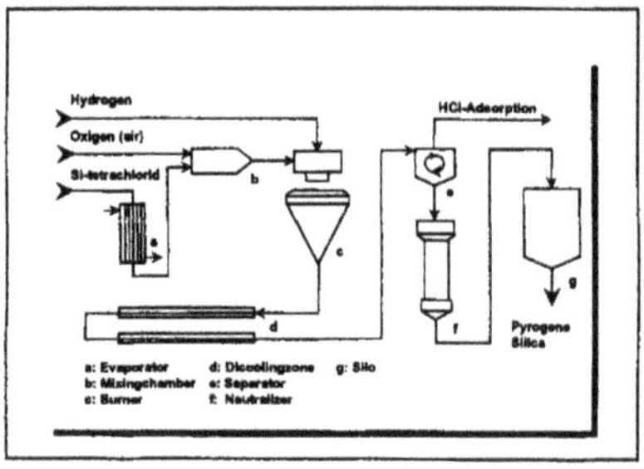

a)

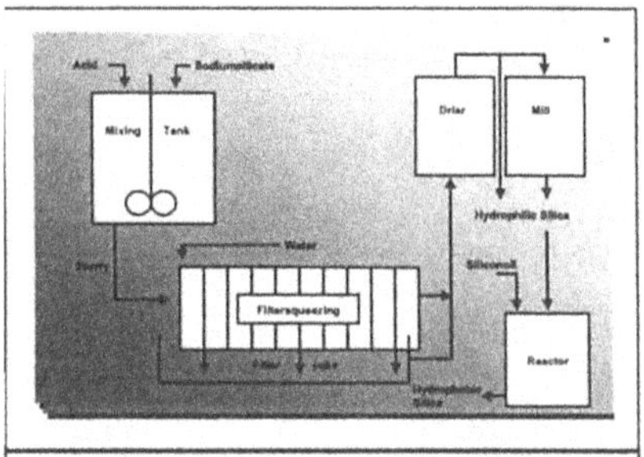

b)

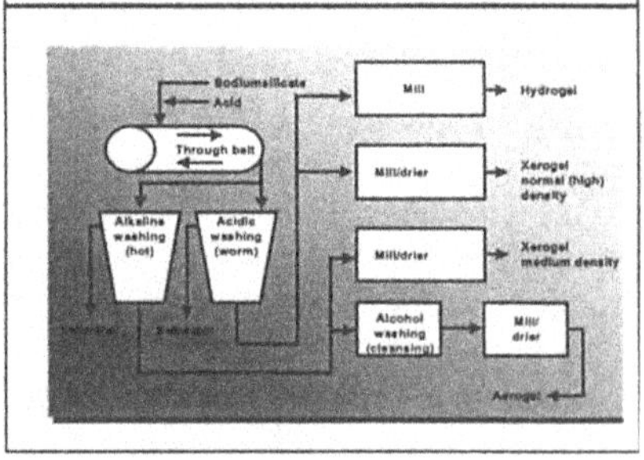

c)

Figure 1
Production Process a) fumed silica,
b) precipitated silicas and c) silica gels.

Coating formulators use matting agents with quite big particles to get dull matt paints and with quite small particles to get a silk matt paint. But it is well known that the degree of mattness, the uniformity and the general optical impression of the matt coatings also depend on the particle size distribution and the morphology of the particles, which can be seen in the kind of surface roughness determined by the Hommel Tester (Figure 2).

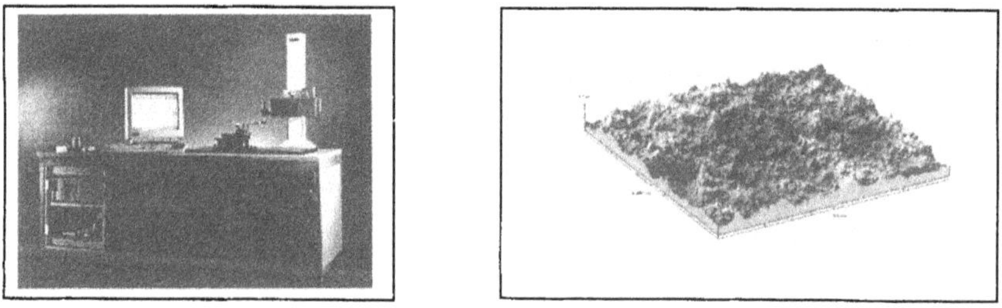

Figure 2   *Hommel-Tester*   *Surface of roughness of a coil coat*

In the past, matting agent manufacturers almost always used SEM or TEM methods to determine the average particle size and the results were given in μm (Figure 3a). However, it is more meaningful to determine the particle size by means of laser diffraction methods. There are a number of suitable devices such as those from Malvern, Cilas or Coulter LS. In addition to the average particle size (D_{50}), these methods also allow the particle size distribution to be determined. However, there is no correlation between the TEM method and laser diffraction measurements (Figure 3b).

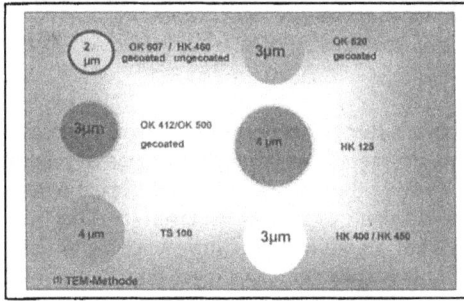

Figure 3   *a) Particle Sizes of various ACEMATT grades*

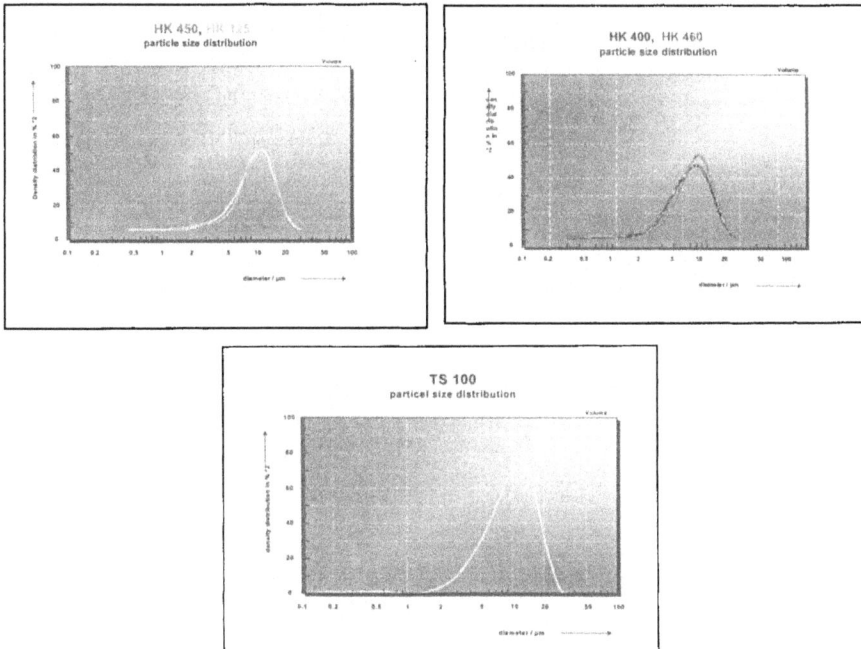

b) Particle size distribution curves of ACEMATT HK 450/HK 125, HK 400/HK 460 and TS 100

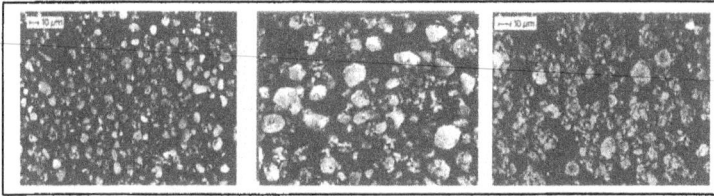

Figure 4   *SEM-Photos of ACEMATT HK 460, HK 450 and TS 100*

There are two main differences between the known, established matting agents and the new generation of matting agents with a higher degree of matting efficiency, affecting the gloss/sheen ratio. Firstly, there are the specific precipitation parameters, which influence the characteristic properties such as BET surface, DBP index and tapped density. Secondly, there is the type of drying and grinding technologies used, which determine the mode of action of the matting agent. In this respect the average particle size, particle size distribution and particle morphology must be considered.

Two new untreated ACEMATT grades were developed through targeted variation and optimisation and a consistent technical implementation of the above-mentioned parameters. The physical and chemical properties of these two matting agents are listed in comparison to ACEMATT TS 100 in Table 1 (Figure 4).

Table 1   *Physical and chemical properties of ACEMATT HK 450, HK 460 and TS 100*

Typical values		HK 450	HK 460	TS 100
Ignition loss	%	5	5	2.5
Drying loss	%	7	7	4
pH-Value		6.5	6.5	6.5
Sulfate content	%	0.8	0.8	1.9
Particle size				
Average value (TEM)	μm	3	2.5	4
d_{50}-values (Laser diffraction method)		10.8	7.3	11.0
Surface treated		--	--	--
Tamped density	g/l	85	85	50
Density	g/cm^3	2	2	2.2
Oil adsorption	g/100 g	255	255	360
SiO_2-content	%	98.5	98.5	99.8

3 APPLICATION

In the following segment, the application properties of these new matting agents will be compared to other significant, three available matting agents.

Since mainly fine particle matting agents are required for furniture coatings, ACEMATT 460 should be mentioned in this regard. To stress once more the specific differences mentioned at the start, Table 2 with comparisons between the physical and chemical properties of ACEMATT HK 460 and commercially competitors' products is introduced.

Table 2 *Comparison of the physical and chemical data of ACEMATT HK 460 and three competitive products*

		HK 460	Competitor 1	Competitor 2	Competitor 3
Drying loss	%	7	7	4	4
pH-value		5-7	3.3	6-8	6-8
Oil absorption	ml/100 g	255	330	250	250
Particle size/					
Coulter counter	µm	n.a.	n.a.	3.3	2.5
Particle size					
Laser diffraction method	µm	7.3	7	6.6	5.3
Particle Size (TEM)	µm	2,5	2,6	3	2.5
Grind gage value	µm	24	26	30	20
Surface treated		no	no	no	yes

To differentiate between the matting efficiency of the matting agents, a formulation was deliberately selected which is not particularly practice oriented, but which provides us with very dissimilar and reproducible results due to its composition, and thus shows the differences in efficiency of the matting agents (Table 3).

Table 3 *Test formulation of the furniture coating*

2pack-PUR-Furniture Coating

		parts by weight
Butyl acetate 98 %		83.0
Ethoxypropyl acetate		165.0
Desmophen 800		150.0
Desmophen 1100		200.0
CAB 381 – 0.5	10 % in Butyl acetate 98 %	30.0
Mowilith 20	50 % in Butyl acetate 98 %	30.0
Baysilone OL 17	10 % in Xylol	1.0
BYK 361		3.0
Xylene		338.0
Total:		1000.0
Viscosity DIN 4	14 – 15 s	
Mixing rate:	2:1 Desmodur L 75	

Apart from matting efficiency, the transparency of the coating is particularly important in colourless furniture coatings and even more so in the case of dark or stained woods. Also in this case, the new matting agent produces some very good results, not just in regard to transparency, with a degree of gloss of 25 at a 60° angle, but also in the small amount of material required to produce this result – which again is evidence of its high degree of matting efficiency (Figure 5).

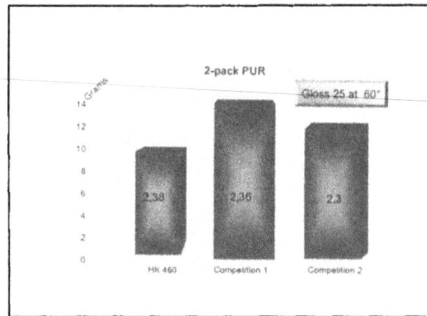

Figure 5   *Gloss of Acematt HK 460 versus competition products (formulation of Table 3)*

However, it is not just transparency and matting efficiency that are important for the appearance and the surface quality of furniture coatings, the chemical resistance must also not be disregarded. Household chemicals were tested in accordance with DIN 68861 Part 1 B. Here; too, ACEMATT HK 460 produced some very good results in comparison to other non-coated matting agents (Table 4). For comparison reasons the surface treated ACEMATT OK 520 was included.

Table 4   *Chemical resistance of various matting agents (formulation of Table 3)*

	HK 460	*Competition 1*	*Competition 2*	*Competition 3*	*OK 520*
Gloss increase					
Coffee	slightly	slightly	slightly	slightly	slightly
Deionised water	slightly	slightly	slightly	slightly	slightly
Red wine	slightly	slightly	slightly	slightly	slightly
Ethanol 48 %	slightly	slightly	slightly	slightly	slightly
Blister					
Coffee	m 1 g 1	m 1 g 1	m 1 g 1	m 1 g 1	m 1 g 1
Deionised water	m 1 g 1	m 1 g 1	m 1 g 1	m 1 g 1	m 1 g 1
Red wine	m 0 g 0	m 0 g 0	m 0 g 0	m 0 g 0	m 0 g 0
Ethanol 48 %	m 0 g 0	delaminated	delaminated	m 0 g 0	m 0 g 0
Paint film defect/					
Delamination					
Coffee	No	No	No	No	No
Deionised water	No	No	No	No	No
Red wine	No	No	No	No	No
Ethanol 48 %	No	Yes	Yes	No	No

Even in waterbased furniture (Table 5) and parquet coatings (Table 6) these matting agents perform in terms of the matting efficiency (Figure 6) very well. It is also interesting to demonstrate the outstanding performance of ACEMATT TS 100 in water based coatings in terms of the matting properties related to the dry film thickness (Figure 7).

Table 5   *Water Based Furniture Coating*

	Parts p. Weight
Neocryl XK 15	790.0
Byk 024	8.0
Coatex BR 125 P / deion, Water 1:1	6.0
Deion, water	89.0
Dowanol DPM	21.0
Dowanol DpnB	21.0
Proglyde DMM	18.0
Byk 346	4.0
Aquacer 513	43.0
Total	1000.0

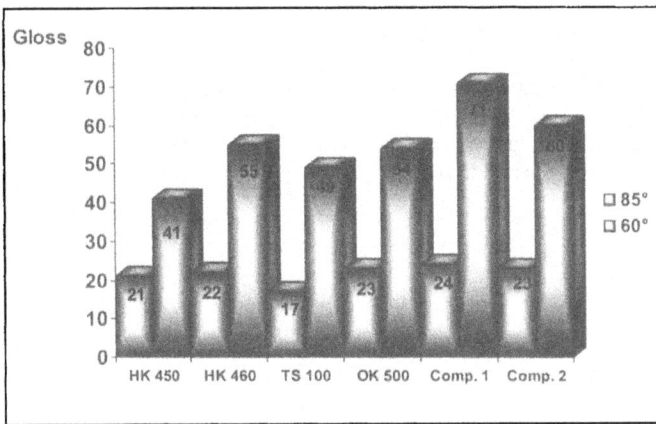

Figure 6   *Matting efficiency of ACEMATT vs. Competition (formulation of Table 5)*

Table 6   *Waterbased parquet coating*

	Parts p. Weight
Neo Pac E 140	923.0
Tego Foamex 822	6.0
Borchigel L 75 N (1:1 in water)	18.0
Viscalex VG 2	3.0
Dowanol DPM	45.0
Tego Glide 410	5.0
Total	1000.0

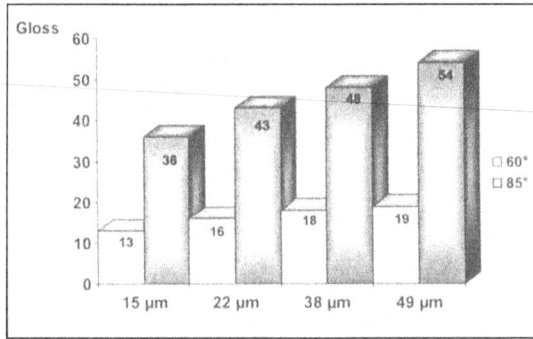

Figure 7   *Gloss versus dry film thickness with ACEMATT TS 100 (based of formulation according Table 6)*

Especially the formulators of thixotropic wood coatings are very often faced with a viscosity drop of the thickened coating when silicas are used to reduce the gloss.

Also here the formulator must carefully select what kind of silica should be used. In the reference coating formulation (Table 6) precipitated silicas perform better than silica gels.

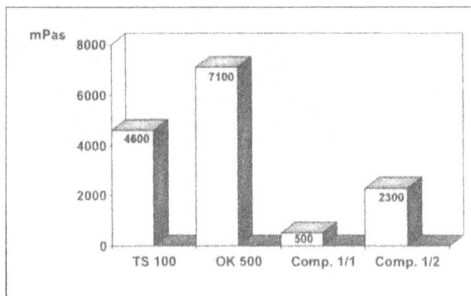

Figure 8   *Viscosity stability Haake RV 6330 (formulation of Table 6)*

4 SUMMARY

To sum up, it can be said that the new generation of untreated silicas Acematt HK 450 and HK 460 set new standards in matting and matting efficiency. But it must be pointed out that these statements are very system-specific. Uncoated matting agents are not generally as universally useable as coated matting agents. But particularly in water-based systems, ACEMATT TS 100 remains the first choice in matting agents due to the sum of its properties.

5 REFERENCES

Technical Bulletin Pigments No. 21 and 32, Company Brochures of Degussa AG/Frankfurt (Main)
EP 0922671, 1997, Degussa AG/Frankfurt (Main)
Ullmann´s, *Encyclopädie der technischen Chemie,* Verlag Urban+Schwarzenberg, Munich

ANTIFOAMS

Hans-Guenther Schulte

Cognis Deutschland GmbH u. Co. KG, Coating Additives, Geb. C 02, Henkelstrasse 67, 40589 Dusseldorf, Germany.

1   INTRODUCTION

In daily life foam is generally a pleasant phenomenon and we feel well in bubble baths, and like to wash our hair with creamy shampoos. We enjoy fluffy cakes with cream and a nice cappuchino with foamed milk – both of these of great importance to the consumer. Many technical applications also rely on a stable foam like manufacturing of isolation materials, flotation of ores or fire fighting.

But in many industrial processes foam is an annoying factor. Cans cannot be filled properly, surfaces show defects, production processes are interrupted or slowed down. Then foam destroying substances, so called defoamers are added. The phenomenon of foam is not limited to water-based systems only. Also solvent-based or solvent-free systems sometimes do have air entrapments and a deaeration is essential for proper performance and appearance. Before taking a look into how foam can be destroyed it is necessary to understand how foam is stabilised, and which substances bring this about - because *pure liquids do not foam!*

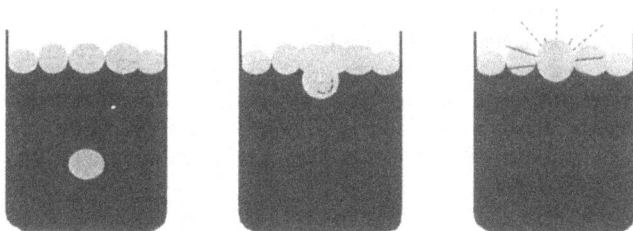

Figure 1 *Pure liquid bubbles collapse*

2   THEORY OF FOAM AND FOAM CONTROL

Foam consists of gas dispersed in a relatively small amount of liquid [1]. Two extremes and a continuum of intermediate stages are possible. An example of one extreme is a freshly poured glass of beer[2] which besides the carbon dioxide already has a significant amount of liquid present. The foam bubbles have no connection to each other and are ball shaped. Because of its much higher specific weight the liquid drains down between the bubbles wile the gas bubbles concentrate on top and by the time become 'dry'. The two stages are therefore called wet or spherical foam and dry or polyhedral foam, see Fig. 2.

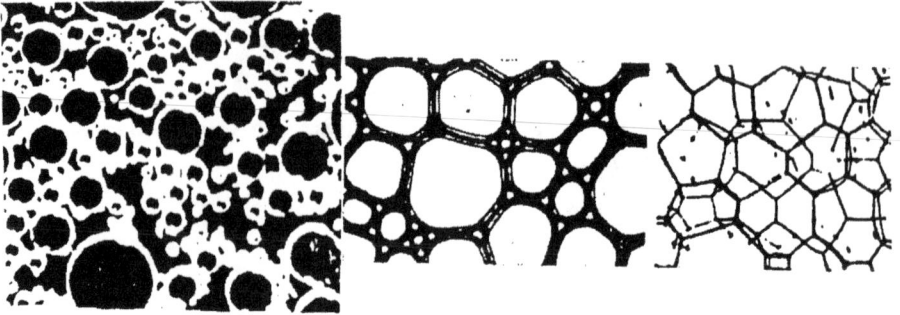

(a)   **(b)**   **(c)**

Figure 2 *(a) Wet (spherical) foam, (b) transition state and (c) dry (polyhedral) foam*

Foams are thermodynamically unstable[3]. The Gibbs equation for a system is given by

$$dG = VdP + SdT + \sigma dA \ (+ \mu_i \, dn_i + \ \ldots \ )$$

In the case where no chemical reaction takes place and T, P are constant, the equation after integration becomes $\Delta G = \sigma \Delta A$ [4].

In simple words: any increase of the surface area (generation of foam) will result in a positive ΔG. This requires work and will never occur spontaneously. Foam, generated by agitation however, can be stable for several hours under well-controlled conditions free of external influences[5] before it collapses.

The situation changes dramatically in the presence of surfactants, which locate themselves at the interface between gas and liquid. They do not change the physicochemical principals but can kinetically stabilise a foam lamella for years under carefully selected undisturbed conditions[6]. These substances consist of a minimum of one hydrophilic and one hydrophobic group. These molecules are called *surfactants*[7] and are often schematically drawn like in fig. 3.

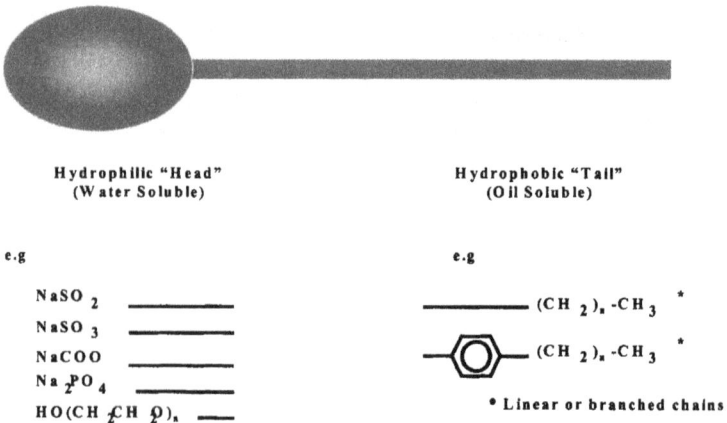

Hydrophilic "Head"   Hydrophobic "Tail"
(Water Soluble)   (Oil Soluble)

e.g   e.g

$NaSO_2$ _____   _____ $(CH_2)_x$-CH_3 *

$NaSO_3$ _____

$NaCOO$ _____ $(CH_2)_x$-CH_3 *

Na_2PO_4 _____

$HO(CH_2CH_2O)_x$ _____   * Linear or branched chains

Figure 3 *Surfactant molecules*

Because of their amphiphilic structure they can dissolve in polar liquids like water and in nonpolar ones like oils. But nonetheless the hydrophobic end of the molecule tries to leave the water and concentrate at the interface to air, which is hydrophobic as well, Fig 4.

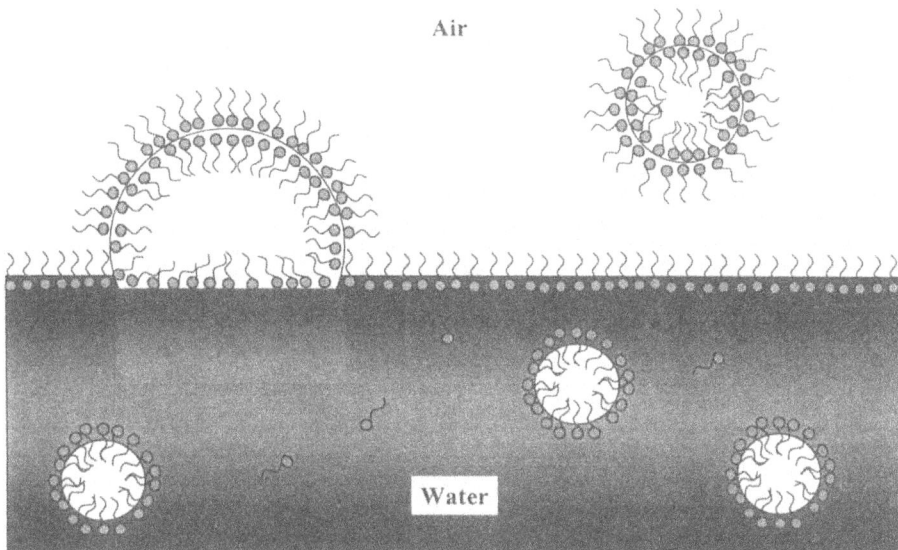

Figure 4 *Formation of a surfactant monolayer on a surface of a liquid and double layers for air bubbles*

Surfactants are often required in systems or in technical processes for various reasons. They can have a variety of structures, either ionic or non-ionic, low molecular weight or polymer, linear or branched. One of their actions in aqueous solution is the stabilisation of foam, which very often makes the use of defoamers unavoidable.

Many of the fundamentals of defoamers were investigated in the middle of the last century. Some basic principles are well understood and will be elaborated further below. However, the huge number of products that exist today is due to the special process requirements they have to fulfil, especially in the coatings field due to optical/aesthetic reasons.

2.1 Physicochemical Fundamentals of Surfaces with regard to Foam and Defoaming

Water is a wonderful nontoxic, non-flammable 'solvent' or process liquid. One of its drawbacks however, is the very high surface tension of 72 dyn/m. Many substrates simply would not be wetted by water and washing surfaces, applying surface coatings and other processes would be impossible. Only by the use of surfactants that reduce the surface tension do many things become possible. The reduction of surface tension by surfactants is shown in fig. 5. To any other surface which approaches or which is approached by the surfactant solution it looks at first like an organic nonpolar liquid. Also, the *contact angle* is much more acute than with pure water alone. Air therefore becomes a much more similar or acceptable medium.

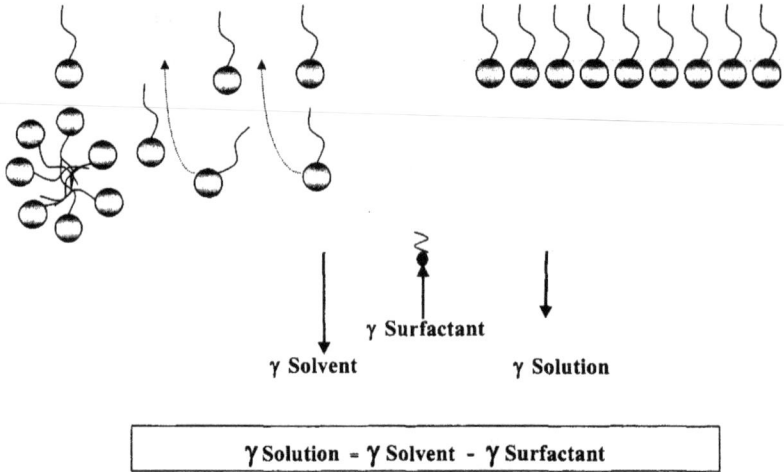

$$\gamma \text{ Solution} = \gamma \text{ Solvent} - \gamma \text{ Surfactant}$$

Figure 5 *Reduction of Surface Tension by Surfactants*[8]

Small droplets do have a higher vapour pressure because of their curvature. The smaller the droplet is, the higher the vapour pressure. The relationship is shown in the LaPlace equation[9], (fig. 6) where the radius is in the denominator. For foam, this means that in very thin lamellae the vapour pressure is higher than in larger volumes e.g. lamellae intersections, see fig. 7.

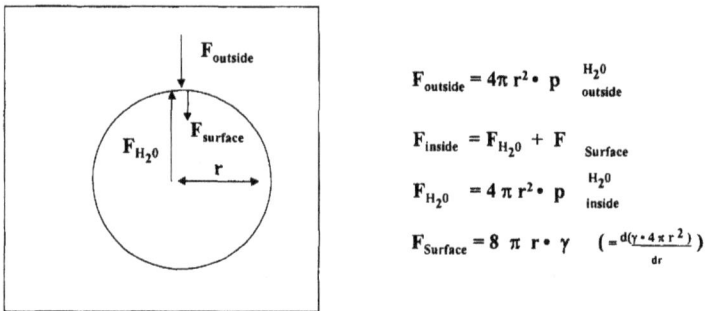

$$F_{outside} = 4\pi r^2 \cdot p^{H_2 0}_{outside}$$

$$F_{inside} = F_{H_2 0} + F_{Surface}$$

$$F_{H_2 0} = 4\pi r^2 \cdot p^{H_2 0}_{inside}$$

$$F_{Surface} = 8\pi r \cdot \gamma \quad \left(= \frac{d(\gamma \cdot 4\pi r^2)}{dr} \right)$$

Equilibrium condition:

$$F_{outside} = F_{inside} \implies P^{H_2 0}_{inside} - P^{H_2 0}_{outside} = \frac{2\gamma}{r}$$

Figure 6 *LaPlace equation for the vapour pressure of liquids and small radii*

At lamella intersections the liquid is more 'bulky'. Tree film pairs come together at an angle of 120°. These zones are called Plateau Boarders (J.J. Plateau, French Chemist 1861). Because of the pressure difference by diameter as well as symmetrical considerations, the vapour pressure is lower than in a lamella which results in a sucking effect towards Plateau zones, helping the drainage of liquid.

Bubble

Bubble

Area of Reduced pressure
'Plateau Boarder'

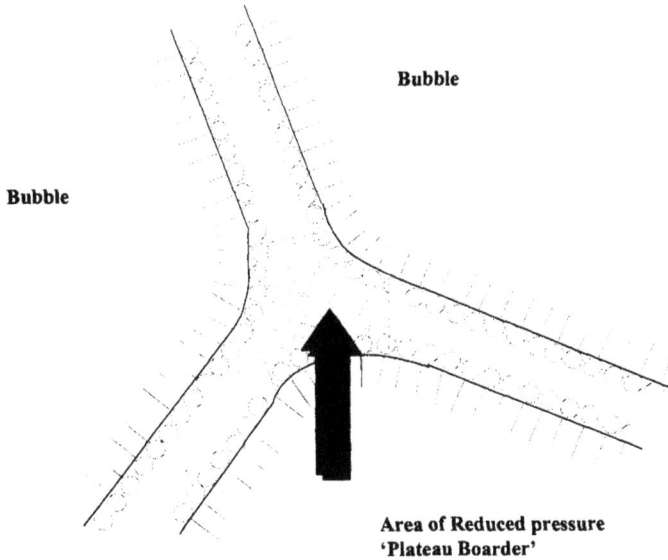

Figure 7 *Plateau Borders at lamella intersections*

Another process in connection with foam is 'flow of a liquid in a capillary'[10].
This process is important during drainage of the liquid. As the radius is influencing the volume per time to the power of 4, flow is strongly reduced at small radii according to HAGEN-POISSEULLES-LAW. (fig. 8)

F l o w o f a l i q u i d i n a c a p i l l a r y

$$\frac{V}{t} = \frac{\pi \cdot \Delta P}{8 \cdot \eta \cdot l} \cdot r^4$$

V = volume ; t = time ; ΔP = pressure difference
η = viscosity ; l = length of capillary ; r = radius of capillary

Figure 8 *Flow of liquids in a capillary*

Low viscous liquids generally have a fast flow and fast drainage, and higher viscosity liquids have a slower flow. This surface flow is also influenced by the presence of surfactants. Surfactants often create a structure in water, e.g. by long chains of polyethyleneoxide. Therefore the drainage is reduced in surfactant double layers. (fig. 9)

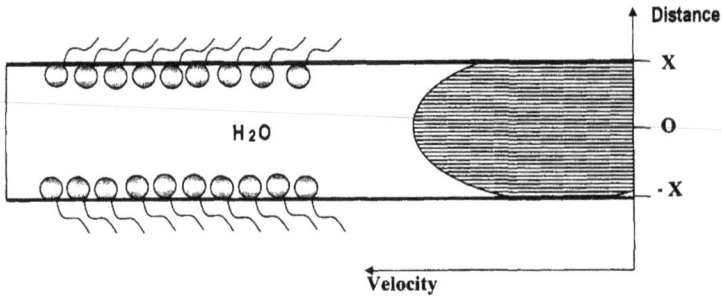

There are two different kinds of foam; fast and slow draining foam

Figure 9 *Influence of surfactants on surface viscosity*

If a double layer is deformed by any kind of stress, which causes a reduction of density of surfactant molecules, this area has a higher surface tension (and higher energy). The system tries to compensate by a) migration of surfactant molecules from the bulk liquid into the area and b) shift of molecules from the double layer into the region. By this means, the system stabilises itself. This self-healing mechanism is also known as the Marangony-Effect[11].

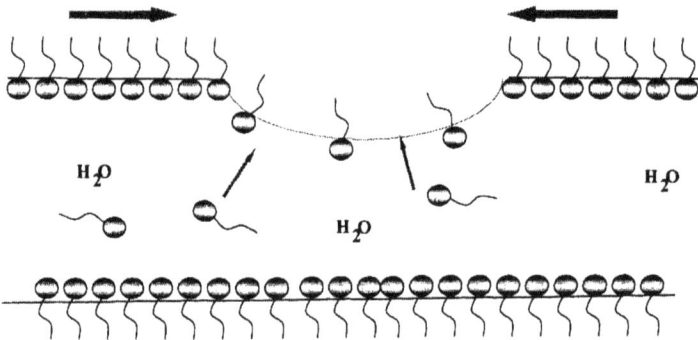

Figure 10 *Marangoni (self-healing) effect in surfactant films*

The last effect to be described here is film elasticity[12]. In case of ionic surfactants the aqueous phase in the double layers contain dissolved counter ions of the surfactants. When the ionic density increases, the repulsive forces of equally charged ions become substantial, see Fig. 11. The repulsive forces are also responsible for a certain elasticity of double layers. The thickness of double layers in the well-known coloured air bubbles lies between 1,000 and 10,000 Å. It can be determined by the order of interferential colours[13]. The process is very dynamic and fluctuates over the surface area. Under certain conditions the drainage reaches an end at a metastable state (so called ´black films´) giving the lamella or bubble a limited time of existence[5].

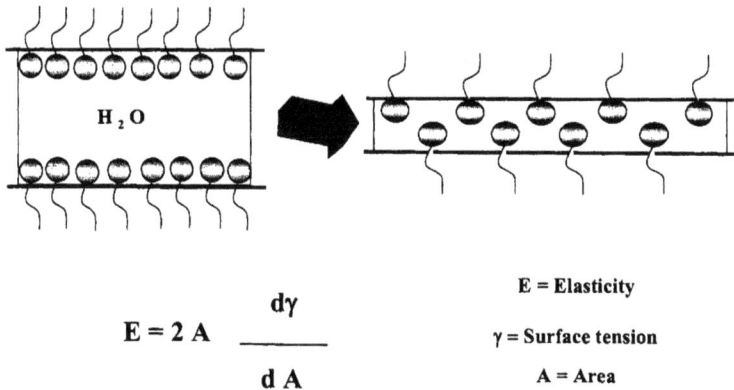

$$E = 2A \, \frac{d\gamma}{dA}$$

E = Elasticity
γ = Surface tension
A = Area

Figure 11 *Film elasticity in surfactant double layers*

The following effects (extracted from the above) have a foam stabilising effect:

- Reduction of surface tension by surfactants
- Surface viscosity of surfactant layers
- Film-elasticity increase by surfactants
- Electrostatic repulsion of electric double layers
- Flow reduction in capillaries
- Marangony effect

Double layer destabilising effects include:-

- Surface tension of solvent (water)
- Sucking effect at junction points (Plateau Boarders)
- Drainage by gravitational forces

The result is a finite lifetime of lamellaes or foam bubbles respectively.

2.2 Theory of Foam Inhibition and Foam destruction

Foam can basically be destroyed either by mechanical action or by the addition of special agents. These agents are called by various names like *Antifoams, Defoamers, Deaerators, Foam Control Agents, Antifoaming Agents* etc. depending on the time or the condition during which the defoamer is active. Although many formulations can do both and there is a continuum of transitional states, we can define two extremes:

- a) an antifoam is mainly active in 'wet foams' accelerating the drainage and improving the coalescing of small bubbles before a stable polyhedral foam is formed. It makes small air bubbles coalesce so that they can rise faster to the surface. It also reduces surface viscosity which further speeds up the deaeration process.
- b) a defoamer destroys stable polyhedral foam by
  a decrease of film elasticity by spreading on double layers
  an increase of surface tension by adsorption of surfactant molecules on adsorbers
  a disturbance of double layer structure by particles.

The instantaneous destruction of polyhedral foam is called "knock-down effect" by some authors.

Most of the defoamer formulations contain hydrophobic particles (for composition see section 3). The contact angle, shape and size of these particles strongly contribute to the defoamer efficiency. Besides adsorbing surfactant molecules the hydrophobic particle reduces the thickness of the lamella by increasing the contact angle.

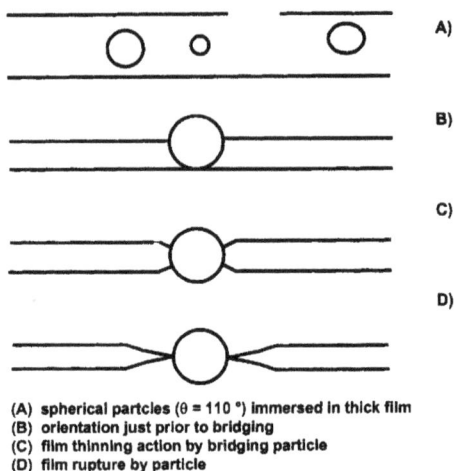

(A) spherical partcles (θ = 110 °) immersed in thick film
(B) orientation just prior to bridging
(C) film thinning action by bridging particle
(D) film rupture by particle

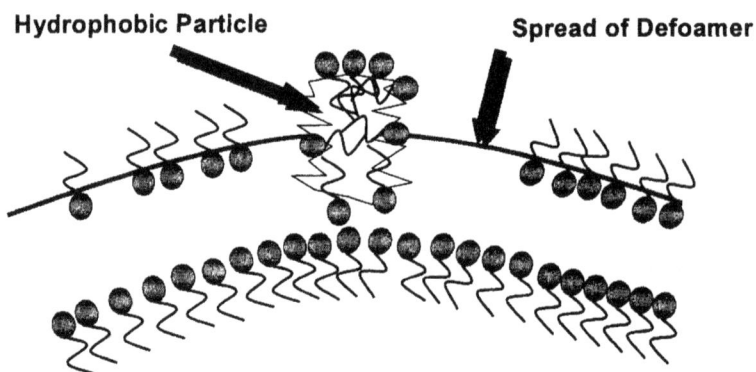

Figure 12   *Film-rupture by dewetting action of hydrophobic parcticle a) and by adsorption of surfactant molecules[14] b)*

But before a defoamer can do its job it has to get from the bulk liquid to the place 'where the bubble is'. At a first step it has to migrate to the surface and as a second it has to be able to spread into the lamellae. The entering process was first described by Robinson and Woods[15] and the spreading process, was described by Harkins[16] and both were combined by Ross [17] in the first comprehensive defoamer theory.
The entering coefficient is shown in fig. 13 and the spreading coefficient in fig. 14

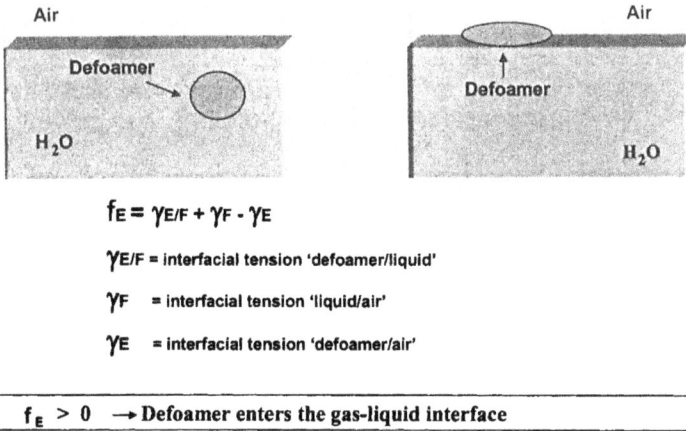

$$f_E = \gamma_{E/F} + \gamma_F - \gamma_E$$

$\gamma_{E/F}$ = interfacial tension 'defoamer/liquid'

γ_F = interfacial tension 'liquid/air'

γ_E = interfacial tension 'defoamer/air'

$f_E > 0 \rightarrow$ **Defoamer enters the gas-liquid interface**

Figure 13 *Entering process and entering coefficient*

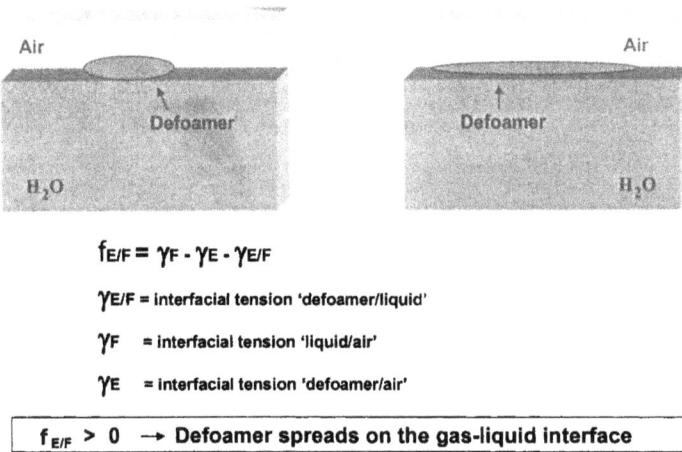

$$f_{E/F} = \gamma_F - \gamma_E - \gamma_{E/F}$$

$\gamma_{E/F}$ = interfacial tension 'defoamer/liquid'

γ_F = interfacial tension 'liquid/air'

γ_E = interfacial tension 'defoamer/air'

$f_{E/F} > 0 \rightarrow$ **Defoamer spreads on the gas-liquid interface**

Figure 14 *Spreading process and spreading coefficient*

For a successful defoaming process ΔG of both steps must be <0 which is the case if the entering and the spreading coefficient (defined by the surface) are positive (>0).

Numerous substances were characterised and tested in their defoaming capability by Ross[18] and others[19]. A correlation of about 90% was achieved which can be viewed as success or disappointment, depending on the standpoint of the observer. Some substances such as phenol and pyridine do not spread on surfactant solutions but can well destroy foam, other exceptions are petroleum ether and olive oil which spread rapidly but have a very poor antifoaming efficiency[20].

2.3 Mechanical Defoaming

Sometimes mechanical defoaming is possible by:
- vacuum dissolvers[21] casing the air bubbles to expand dramatically
- centrifugation
- blowing a jet of air onto the foam[22]
- spraying fine water droplets on foam[23]
- use of shockwaves by ultrasound[24]
- or stirring the liquid with stirrers of various shapes and sizes.

3   COMPOSITIONS OF DEFOAMERS

The following table gives an overview about the most common raw materials used in oil-based formulations including their function[25]:

Mineral oil	Carrier, Spreading Agent
Oxo bottoms	Carrier, Spreading Agent
Alkylbenzene bottoms	Carrier, Spreading Agent
Vegetable oils	Carrier, Spreading Agent
Silicone oils	Carrier, Spreading Agent
Waxes, Soaps, Silica	Adsorption of surfactant molecules
Silicone oils	Spreading agents, adsorption of surfactant molecules, decrease of surface viscosity
Polyalkylene oxides	Spreading agents, adsorption of surfactant molecules, decrease of surface viscosity
Amphiphilic substances like nonionic Surfact.	Assist incorporation, and adjustment of compatibility

3.1 Mineral oil defoamers

Volume-wise, mineral oil based defoamers play by far the most important role world wide[26]. Even though mineral oil has a certain defoaming capability it is generally considered as carrier for more active ingredients. The influence of the defoaming action of the mineral carrier oil is not negligible, given the variety of composition of mineral oils, the differences in cracking and refining processes, and the demand for listing in chemical directories or pharmacopoeia, and limit settings in composition for ecological or toxicological reasons. The highest class of a mineral oil is a paraffinic white oil which is fully hydrogenated, has no odour and fulfils the requirements of Pharmakopeia (DAB)[27].

3.2 Emulsion defoamers

For cost reasons and for the ease of incorporation, mineral oil defoamers can be emulsified *in water to give* an oil-in-water (O/W) emulsion defoamer[28]. The droplet size plays an important role in their efficacy and persistency[29]. One of the difficulties with this approach concerns preparation of a stable emulsion which when diluted becomes unstable in order to drive the active antifoam ingredient to the air/water interface and not remain in the bulk of the organic phase. Inverted water-in oil (W/O) emulsions may be prepared to overcome such difficulties[30]. Microemulsions are claimed to be more stable than macroemulsions and even conventional defoamer compositions, which do not separate on standing, show suitable low viscosities, and can easily be incorporated in latex paints[31].

3.3 Powder defoamers

Defoamer formulations or active ingredients can also be applied on solid carriers like silica or organic substances to yield free flowing powders[32]. They are used in powder type formulations like plasters, mortars, self-levelling flooring components, etc.

3.4 Silicone based defoamers

Silicone oils[33] are particularly effective antifoaming agents because of their low surface tension, thermal stability, chemical inertness, and total water insolubility. However, on their own they show little or no foam inhibiting properties. It is only when they are combined with finely divided silica particles that an effective foam inhibitor is produced. Traditional silicone antifoam compounds are proprietary combinations of polydimethylsiloxanes (PDMS) and such finely dispersed particles of silica gels, pyrogenic or precipitated silicas[34]. These silica particles need an in-situ heat treatment to reach the so-called β-type required for an efficient antifoam activity[35]. By this treatment part of the silicone polymer is bound to the silica filler surface, the hydrophilic silica filler becoming hydrophobic[36]. This has an important effect on the rheological behaviour. The viscosity of the compounds is Newtonian but significantly larger than the theoretical prediction based on the volumic fraction of the spherical particles.

The very low surface tension of the highly branched silicone fluids allows spontaneous spreading over the surface of most surfactant-containing solutions, exposing the hydrophobic silica particles to the interface; in other words, the hydrophobic silica particles act as foam breakers and the hydrophobic silicone fluid is their means of transport to the water/air interface. Polydimethylsiloxanes also traditionally find some use in small concentrations, as components in mineral oil defoamers[37], or they are used as silicone oil emulsions to inhibit foam in aqueous systems. For paint applications, silicone oils generally are too incompatible. They tend to give strong cratering, intercoat adhesion failure and "fish eyes". Even the common complaint that silicones and silicone defoamers contaminate every piece of equipment they come in contact with has validity[38]. To this end, investigators have studied the effect of introducing organic moieties into the polymeric siloxane matrix. Watersoluble and – even better – water-insoluble ABA, A-[B-A]$_z$ or R^1-O-A[B-A]$_m$R^1 polyoxyalkylen-polysiloxane-block-copolymers have been described for direct use or as part of defoamer formulations[39] The main applications are high-gloss and clear coats. Also, in pigment concentrates with very high concentrations of surfactants (up to 25%) they have proved to be efficient. Having similar refractive indexes and good compatibility makes them 'invisible' in paint films.

3.5 Block-Copolymer defoamers

More recent investigations have shown that organically modified polysiloxanes can be replaced by hydrocarbon based blockcopolymers, which – just like modified PDMS – can be water-soluble or water-insoluble. These blockcopolymers are the active ingredients in modern silicone-free and mineral oil-free paint defoamers and also in emission-free (VOC-free) eco-defoamers[40]. Hyper-branched blockcopolymer can be compounded with other defoaming components to yield defoaming results unachievable with conventional products[41].

4   DEFOAMER TESTING AND SELECTION

General, more scientific defoamer tests which concentrate on basic functions are reported in the literature[19]. In technical processes additional requirements led to a number of specific tests being used in this particular industry, e.g. optical requirements in coatings. The type of test as well as the incorporation or application of the substance was found to have a significant influence on the test results[20].

A strong influence on test results is also given by the compatibility of the defoamer. All defoamers regardless of their fundamental chemistry can be divided into three classes[42].

Table 1 *Classification of Defoamers*

Class	Advantage	Disadvantage
Non-emulsifiable	Excellent efficacy and long term persistence	Needs shear for incorporation, poor knock-down effect, limited compatibility
Medium-emulsifiable	Universal, good compromise between immediate action and long term persistence,	Sometimes not sufficiently spontaneous, sometimes efficiency under shear is somewhat reduced
Emulsifiable	Spontaneous activity (knock-down), excellent compatibility, universal, easy to incorporate especially in low viscous systems, used also in latex production	Reduced long term efficiency, limited performance under constant high shear conditions

According to the table above the products to be screened need to be selected first in view of depending on according to the final application. Defoamer tests used in the coating field can be divided into screening or short term tests and more specific tests which often simulate the specific application. Screening tests are often done by using the latex only, but the results obtained here have only a limited validity in the final paint formulation. The following tests must always be carried out alongside comparison tests to a reference, such as a paint with or without a given defoamer.

- Shake test

A well-established simple test is carried out by shaking the polymer dispersion or paints in a closed cylinder and measuring the generated foam in centimeters as well as noting the time for complete foam break down. Higher viscosity material needs to be diluted with water. This test however, is limited to non-pigmented systems.

- Density test

This test can be used for ready paints as well as for binders. Also, here the paint or binder often needs dilution with water, which increases the foaming tendency. A certain volume of the paint is stirred vigorously under defined conditions (time, temperature, stirring speed, geometry of vessel). Then it is immediately transferred to a pyknometer and the weight measured. The higher the density, the more effective is the defoamer. It is also very informative to make an immediate draw-down on a glass or plastic substrate and check whether the defoamer can destroy the foam before the film is cured, and the influence on optical appearance and compatibility can be rated.

The most efficient types should be selected for special application tests. These tests try to mimic the final application.

- Roller test

A small volume of paint is spread by felt or sponge roller on a (preferably closed) surface like glass, plastic film or sealed hard board. After drying the appearance of the film is evaluated. Transparent surfaces can be assessed over a lamp and then be checked for small holes that are due to very small air bubbles, called microfoam. Also, the influence of the defoamer on levelling and gloss can be rated here. Some modern defoamers are very active in destroying foam bubbles very quickly like the newly introduced *FoamStar*® products[41]. Here the bubble break time in seconds is an indication of effectiveness.

- Airless spray test

This application involves, for aqueous coatings, a very intensive microfoam generating process. It can be simulated in the laboratory by using electrical hand pistols which give quite comparable results. Microfoam causes very small defects, which if not eliminated, are a starting point for corrosion. These pin-holes canbest be seen when the panel is held over a strong light where they appear as little stars shining through the film.

- Curtain Coating

This application method is often found in wood and paper coating. Here, defoamers must be very shear stable and not loose efficiency during continuous pumping of the paint. On the other hand the defoamer must not cause disruption of the curtain by incompatibility at any time.

- Automatic testing

The market provides an automatic defoamer testing machine which measures the foam height by photo cells as a function of time. Foam is induced by either vigorous stirring, blowing air through the substance from the bottom through a fritte, or by circular pumping with splashing the liquid back into the cylinder[43].

With these more application related tests, the dosage has to be optimised to get a maximum of defoaming but a minimum of side effects and cost. General concentration curves are shown in Fig. 15. Because crossovers are rarely found it is generally sufficient to carry out screening tests at one concentration.

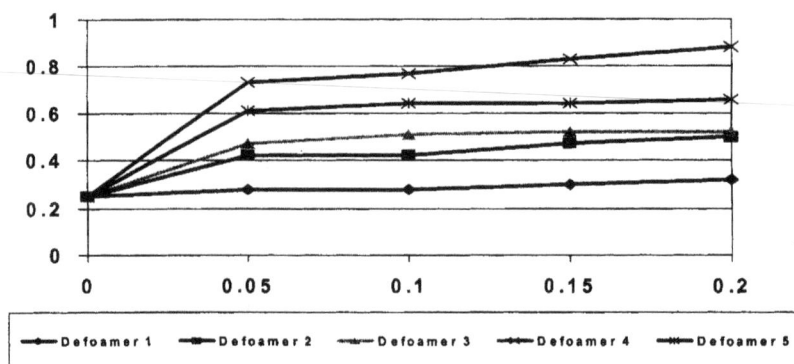

Figure 15 *Defoaming efficiency versus concentration of various defoamers*

Besides checking the efficiency in these tests against air bubbles other film- or surface-effects can be evaluated here such as craters, haze, orange shell, intercoat adhesion (after a second coating) which lead to an exclusion of certain products.

After finishing the first test series the paint or ink including defoamer needs to be stored at elevated temperatures (e.g. 40 or 50°C) for 2-4 weeks that simulates a storage time of about half a year. This is necessary to evaluate the expected defoamer efficacy after storage of the paint or ink. Some defoamers loose part of their function by e.g. migrating into binder particles or other mechanisms. The incorporation of a defoamer is also very important. Too strong a shear can create too small particles that have little effect (slow entering). Too little shear, which often happens in clear coats or other low viscosity formulations, leads to large defoamer particles, which can lead to craters.

Besides initial and long term efficiency and optical properties several other requirements are part of a defoamer selection, such as food contact approval, price, odour, ease of application, universality, no separation upon storage. Some companies try to limit the number of different defoamers e.g. one defoamer for interior coatings, one for exterior and one for high-gloss coatings. The selected defoamers must be very universal and this is only possible if the group of products is relatively similar, e.g. based on one latex and one dispersing agent.

Defoamer selection procedure
1. select defoamer test (screening test and test closely related to application)
2. select defoamers according to compatibility, price, composition, odour, VOC....
3. run screening tests
4. run specific test with different concentrations, usually from 0.1 to 1%
5. run storage tests with best performing products
6. repeat specific tests and select best performing product

5   REFERENCES

1) J.J. Bikermann, *Foams*, Reinhold, New York (1953)
2) R. T. Roberts, *Some aspects of the stability of beer foam*, in R. J. Akers, Ed., Foams, Proc. Symposium Soc. Chem Ind., Brunel University, Academic Press, London, New York, San Francisco (1976)
3) J.H.Aubert, A.M. Kraynik, P.B. Rand, Spektrum der Wissenschaft 7, 126

(1986)

4) Ross, S., *Mechanism of Foam Stabilisation and Antifoaming Action*, Chem.Engineering Progress, 63(9), 41 (1967)

5) Lange, H., *Schäume und ihre Stabilität*, VDI-Berichte 182, 71 (1972)

6) I. Dewar, Proc. Roy. Inst. 22, 179 (1917)

7) H. Stache, K. Koswig, Tensid-Taschenbuch, Hauser, München, Wien (1990)

8) W.D. Harkins, H.F. Jordan, *A Method for the Determination of Surface and Interfacial Tension from the Maximum Pull on a Ring*, J.Am.Chem. Soc. 52, 1751 (1930)

9) W.J. Moore, *Physical Chemistry*, Prentice Hall Inc., Englewood Cliffs, NJ, USA (1972)

10) Tschakert, H.E., Schaum, ein anwendungstechnisches Problem, Tenside 3[9], 317 (1966)

11) Marangoni,C., Nuovo Cimento 3(97), 193 (1878)

12) Gibbs, J.W. *Collected Works*, Vol. I, Longmans, Green & Co. New York 1928 p. 300

13) Miles, G.D., Shedlovsky, L., Ross, J., *Foam Drainage*, J.Phys. Chem. 49, 93 (1945)

14) Rrye, G.C., Berg, J.C., *Antifoam Action by Solid Particles*, J. Coll. Interface Sci. 127(1), 222 (1989)

15) Robinson, J.W., Woods, W.W., *A Method of Selecting Foam Inhibitors*, J. Soc.Chem. Ind. 67, 361 (1948)

16) Harkins, W.D., J. Chem. Phys. 9, 552 (1941)

17) I.D. Morrison, G. Nishioka, Professor Sydney Ross-A Festschrift, Colloid Surf. A 118(3) R3-R4 (1996)

18) Ross, S., Young, G.J., *Action of Antifoaming Agents of Optimum Concentrations*, Ind. Eng. Chem. 43(11), 2520 (1951)

19) Tsuge, H., Ushida, J., Hibino, S., *Measurement of Film-Breaking Ability of Antifoaming Agents*, J.Coll. Interf. Sci. 100(1), 175 (1984)

20) Okazaki, S., Sasaki, T., *Über Schaumverhindern und Schaumbrechen*, Tenside 3(4), 115 (1966)

21) N.N., Coatings World, July/August, 16 (1999)

22) Société des Forges Et Ateliers du Creuzot, Ger. Offen. 2 051 526; Chem.Abstr. 75, No. 7822 (1971)

23) F. Ohl, Chemische Rundschau 4, 271 (1951)

24) F.R. Adams, Tappi 41(5), 173 A (1958)

25) R. Höfer et al., *Foams and Foam Control*, in Ullmann's Encyclopedia of Ind. Chemistry, 5th Ed., A 11, 465, VCH, Weinheim (1988)

26) H.-F. Fink, G.Koerner, Schaumbekämpfung in Ullmanns Encyclopädie der Techn. Chemie 4.Ed.20 page 441

27) Europäisches Arzneibuch, 3. Ausgabe (1997); US FDA CFR § 178.3620a

28) Montani, R.C., Boylan,F.J., Ann. Meet.-Techn. Assoc. Pulp Pap. Ind. 1980, 293 (1983)

29) E. Wallhorn, W. Heilen, S. Silber, *Entschäumer – alles nur Empirie ?*, Farbe & Lack 102 (12), 30 (1996).

30) W. T. Schmidt, C. T. Gammon, *Entschäumerzusammensetzung auf Wasserbasis*, DE 29 44 604, 1979 (Diamond Shamrock)

31) C. T. Gammon, *Microemulsion defoamer composition and a process for producing the composition*, EP 7 056, 1978 (Diamond Shamrock)

32) M. D. Kellert, I. A. Lichtman, *Antischaummittel*, DE 2 048 847, 1970 (Diamond

Shamrock)

33) W. Noll, *Chemie und Technologie der Silikone*, VCH, Weinheim (1968); W. Noll, *Chemistry and Technology of Silicones,* Academic Press, New York (1968)

34) R. E. Patterson, *Hydrophobic Silica Formulations as Antifoams*, Text. Chem. Color 17, 181 (9/1985)

35) S. Ross, G. Nishioka, *Monolayer Studies of Silica/Polydimethyl-Siloxane Dispersions*, J. Colloid & Interf. Sci., 65, 216 (1978)

36) A. Pouchelon, C. Araud, *Méchanismes d'action des antimousses silicones,* J. Chim. Phys. 93, 536 (1996); A. Pouchelon, C. Araud, *Silicone Defoamers*, J. Dispersion Sci. Technol., 14 (4) 447 (1993)

37) F. J. Leonard, A. De Castro, T. F. Groll Jr., *Defoaming Composition*, US 2 843 551, 1954 (Nopco); R. D. Kulkarni, E. D. Goddard, M. P. Aronson, *Hydrocarbon oil based silicone antifoams*, EP 121 210, 1984 (UCC)

38) R. T. Maher, E. M. Antonucci, *A Formulator's Guide to Silicone Defoamers*, Mod. Paint Coatings, 43 (Dec. 1979)

39) D. L. Bailey, *Verfahren zur Herstellung von Polysiloxan-Polyoxyalkylen-Blockmischpolymerisaten*, DE 10 12 602, 1955 (UCC); H.-F. Fink, H. Fritsch, G. Koerner, G. Rossmy, G. Schmidt, *Zubereitung zur Entschäumung*, DE 24 43 853, 1974 (Goldschmidt);); H.-F. Fink, H. Fritsch, G. Koerner, G. Rossmy, G. Schmidt, *Defoaming composition and method of defoaming aqueous solutions or dispersions*, US 4 028 218, 1974 (Goldschmidt); M. Grünert, H. Müller, H. Tesmann, H.-U. Hempel, *Verwendung einer Polysiloxan-Zubereitung als Entschäumer in wäßrigen Dispersionen und Lösungen von Kunstharzen*, DE 31 23 103, 1981 (Henkel); R. Berger, H.-F. Fink, O. Klocker, R. Sucker, *Verfahren zum Entschäumen wässriger Dispersionen polymerer organischer Substanzen durch Zusatz von Polyoxyalkylen-Polysiloxan-Blockmischpolymerisaten*, EP 331 952, 1989 (Goldschmidt)

40) Cognis Geschäftsbericht; H.-G. Schulte, EO/PO-Blockcopolymers, XXVth Int. Conf. In Org. Coatings, Vouliagmeni (Greece) 1999; J. Schmitz, H.-G. Schulte, M. Fies, R. Höfer, *Optimal formulieren mit variablen Bauseinen*, Farbe + Lack 106(3) 28 (2000).

41) K. Breindel, *Molecular-based defoamer eliminates foam and enhances coating properties*, Modern Paint & Coatings, (May 2000)

42) W. Gress, *Schaumbildung und Schaumkontrolle in wässrigen und lösemittelhaltigen Lacken*, Swiss Chem, 14[a], 59 (1992

43) Sita Messtechnik GmbH, Dresden (Germany)

SILICONE SURFACE ACTIVE AGENTS

Donna Perry

Dow Corning Ltd, Cardiff Road, Barry, S. Glamorgan, CF 63 2YL

1   INTRODUCTION

Surface-active agents are chemicals that reduce the surface energy of a material to which they are added by absorption at the air/liquid interface. The increasing popularity of waterborne coatings presents a continuing challenge for formulators. There is the need to formulate more environmentally acceptable products to comply with government regulations on VOC's, to reduce the flammability and to generally reduce the hazard profile of coating systems. In comparison with solvent-borne systems, water-borne coatings exhibit a much higher surface tension. This results from water having a surface tension of 72.4 mN/m. This results in water borne coatings exhibiting poor substrate wetting capabilities when compared to solvent based equivalents. The high surface tension of water is due to its polarity and the cohesive hydrogen bonding between the water molecules. Additives or Surface-active agents are used to modify the surface properties of water-borne coatings in order to achieve the desired flow, levelling and wetting performance.

2   SURFACE TENSION EFFECTS IN WATER-BORNE COATINGS

Surface tension can be considered in various ways, but perhaps best as a result of the forces of attraction existing between the molecules of a liquid. It is measured by the force per unit length acting in the surface at right angles to an element of any line drawn in the surface (mN/m).

Wetting is the effect of displacing one fluid from the substrate surface by another liquid. In general, the liquid must have the same or a lower surface tension than the substrate otherwise no wetting will be achieved. This will produce a positive spreading coefficient and lead to a good coating. This can be best expressed in the equation (1) shown. By addition of the correct additive, the surface tension of the coating can be modified to improve substrate wetting. The additive acts by lowering the surface tension of the coating enabling it to wet out over the substrate.

Levelling is the process of eliminating the surface irregularities of a continuous liquid film under the influence of the liquid's surface tension. Levelling is an important step in obtaining a smooth and uniform film. Concentration and surface tension gradients in coatings are the main causes of poor levelling which develop as the coating dries due to variations in drying rates. Addition of the appropriate additive will modify the surface tension. For instance, a silicone additive will migrate to the liquid/air interface giving an even surface tension across the film during the drying process thus reducing surface tension gradients.

Wetting and levelling are therefore totally different processes. Though the surface tension of the coating plays a very important role in both processes, the effects of surface tension on spreading is opposite to that observed in levelling.

Substrate wetting can be subdivided into two distinct categories of Spreading wetting and Adhesion wetting. Spreading wetting is necessary for even coverage of applied films, in particular for the low surface energy substrates of many plastics.

Adhesion wetting is defined in equation (2). It is important e.g. in certain printing processes in which it would not be desirable for the applied ink to spread across a surface and then merge with other colours.

Spreading wetting:

$$S_{(Spreading\ coeff.)} = -\ G_{Spreading}\ /\ A$$
$$= \gamma_{Substrate} - \gamma_{Substrate/Liquid} - \gamma_{Liquid}$$
$$= \gamma_{Liquid}(\cos\theta\ -1) \tag{1}$$

When $\gamma_{Substrate/Liquid} > 0$, spreading will occur spontaneously.

Adhesion wetting:

$$W_{A(Work\ of\ adhesion)} = -\ G_{Adhesion}\ /\ A$$
$$= \gamma_{Substrate} + \gamma_{Liquid} - \gamma_{Substrate/Liquid}$$
$$= \gamma_{Liquid}(\cos\theta\ +1) \tag{2}$$

(γ (gamma) = Surface tension)
(θ (theta) = Contact angle)

3   SILICONES AS SURFACE ACTIVE AGENTS

Of the various surface active chemistries currently available, this paper will mainly concentrate on a class of materials called Silicone Polyethers. This family of copolymers is used to provide multifunctional benefits in water borne systems. The main uses of silicone polyethers in inks and coatings include de-foaming, de-aerating, improved substrate wetting, levelling and enhanced slip properties (1,2). The three most common molecular structures for silicone surfactants are rake type copolymers, ABA copolymers and trisiloxane surfactants. These are illustrated in Figs 1,2 and 3 respectively and the performance of these structures will be described in two types of coatings:

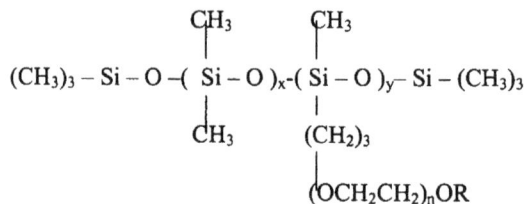

$$(CH_3)_3 - Si - O - (\underset{\underset{CH_3}{|}}{\overset{\overset{CH_3}{|}}{Si}} - O)_x - (\underset{\underset{(CH_2)_3}{|}}{\overset{\overset{CH_3}{|}}{Si}} - O)_y - Si - (CH_3)_3$$

$$(OCH_2CH_2)_n OR$$

Figure 1 *Molecular Structure of a Rake type silicone surfactant*

$$RO\,(CH_2CH_2O)_n\,(CH_2)_3 - \underset{\underset{CH_3}{|}}{\overset{\overset{CH_3}{|}}{Si}} - O - (\underset{\underset{CH_3}{|}}{\overset{\overset{CH_3}{|}}{Si}} - O\,)_x - \underset{\underset{CH_3}{|}}{\overset{\overset{CH_3}{|}}{Si}} - (CH_2)_3(OCH_2CH_2)_nOR$$

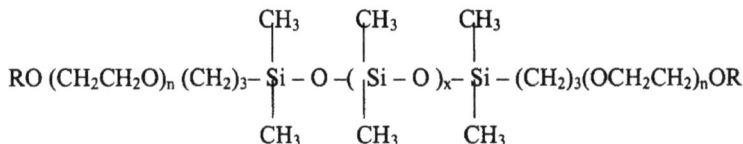

Figure 2 *Molecular Structure of an ABA type siloxane surfactant*

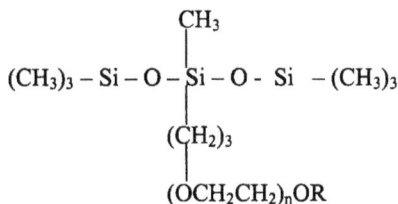

$$(CH_3)_3 - Si - O - \underset{\underset{(OCH_2CH_2)_nOR}{|}}{\overset{\overset{CH_3}{|}}{\underset{|}{Si}}} - O - Si - (CH_3)_3$$
$$(CH_2)_3$$

Figure 3 *Molecular Structure of a trisiloxane surfactant*

Silicones are highly surface active due to their low surface tension caused by the large number of methyl groups and due to the small intermolecular attractions between the siloxane hydrophobes. The siloxane backbone of the molecule is highly flexible which allows for maximum orientation of the attached groups at interfaces .

Silicone polyethers are non-ionic in nature, and have both a hydrophilic part (low molecular weight polymer of ethylene oxide or propylene oxide or both) and a hydrophobic part (the methylated siloxane moiety). The polyether groups are either ethylene oxide or propylene oxide, and are attached to a side chain of the siloxane backbone through a hydrosilylation or condensation process. They can form a rake-like, comb structure, or linear structure. Silicone polyethers are stable up to 160-180 degrees celsius. There is a great degree of flexibility in designing these types of polymers. A very wide variety of co-polymers is possible when the two chemistries are combined.

They can be varied by molecular weight, molecular structure (pendant/linear) and the composition of the polyether chain (EO/PO) and the ratio of siloxane to polyether. The molecular weight influences the rate of migration to the interface. The increased molecular weight of a polymer typically leads to an increased viscosity, but may also give greater durability to surfaces and improved shine level. Other variables include absence or presence of functionality or end groups on the polyether fragments.

Depending on the ratio of ethylene oxide to propylene oxide these molecules can be water soluble, dispersible or insoluble; obviously, for efficient wetting, these surfactants need to have good solubility in solution. As surfactants, they have the ability to produce and stabilise foam depending on their structure. Conversely, if their solubility parameters are low then they behave as anti-foaming agents.

The different structures affect how the molecules can pack at an interface. With pendant types in aqueous media, the silicone backbone aligns itself with the interface leaving the polyalkylene oxide groups projecting into the water. Linear types are considered to form a very flattened "W" alignment where the central silicone portion of the molecule aligns with the interface and the terminal groups are in the aqueous phase. The amounts to be added vary between 0.01% and 0.5% on the total formulation.

This paper will now discuss the behaviour of silicone polyethers in water, together with some studies in coating formulations. The results are compared with other additives that are used in water-borne coatings such as fluorocarbons and acetylenic glycols.

4   SILICONE POLYETHERS AS WETTING AGENTS

In this study the three different siloxane surfactants described above were evaluated in aqueous solution, in a water based printing ink, and in a water based polyester coating in comparison with fluorosurfactants and acetylenic glycols, which are also used as wetting agents. To achieve good wetting and a positive spreading coefficient, the surface tension of the coating must be lower than the critical surface tension of the substrate.

Table 1 *Equilibrium surface tension with a Krüss K10T tensiometer and a platinum Wilhelmy plate, additive concentrations 0.1%*

Surfactant	Equilibrium Surface tension(nM/m)	CMC (%)
Trisiloxane A	20.5	0.008
ABA Siloxane B	29.0	0.025
Rake Siloxane C	29.9	0.018
Fluorosurfactant	17.5	0.030
Ethoxylated Acetylenic Diol	25.3	0.050

Water has a typical surface tension of 72mN/m and, as can be seen from the above table, all the surfactants tested reduced the surface tension of the system, and as a result the aqueous medium wets more efficiently. For the silicone surfactants the best results were achieved with product A, a low molecular weight material. Trisiloxane A gives a very low figure that is only improved upon by the fluorosurfactant. The Critical Micelle Concentration (CMC) is the required level of product to initiate the formation of micelles in the bulk of the liquid. Up to this point the surfactant added to the water migrates to the liquid/air interface to form a film which reduces the surface tension. The low CMC for A shows its high packing efficiency at the interface, in the much lower level required in comparison to the other products.

In this context it is important to understand the difference between equilibrium and dynamic surface tension. For an equilibrium surface tension measurement, a platinum plate is immersed in the test solution and then slowly withdrawn. The force required to remove the plate from the solution is a measure of the surface tension of that liquid. Dynamic surface tension can be measured by an instrument that bubbles air through the test liquid at an increasing rate, during which the maximum pressure that is required to form a bubble is measured. As the bubble rate increases from 1 bubble per second to 10, the time to create the new interface (liquid/air) decreases. This is effectively measuring how quickly the surfactant lowers the surface tension. The dynamic method is more representative of the coating application e.g. spraying. Under these circumstances it is important to know how quickly a surfactant can migrate to newly formed interfaces. An ideal surface active agent would provide excellent surface tension reduction under both equilibrium and dynamic conditions.

Fig 4 shows dynamic surface tension results for the above materials at 0.1% in water. Trisiloxane A gives better performance than the other silicone polyethers, but does not give the very low values achieved by either the fluorosurfactants or ethoxylated acetylenic glycols. However, it must be remembered that dymanic surface tension only represents one aspect of the total phenomenon.

Dynamic surface tension in water

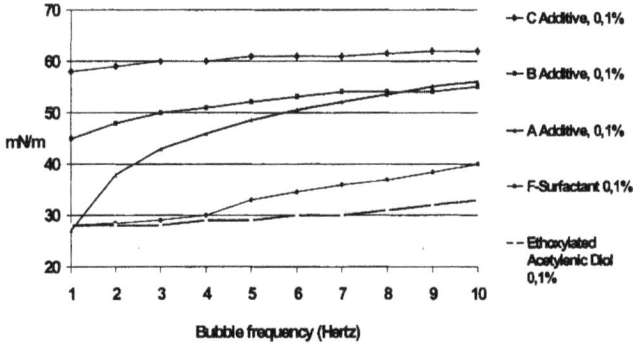

Figure 4 *Dynamic Surface tension,maximum bubble pressure(Kruss BP1),0.1% in water.*

In practice, not only the conditions at the air/liquid interface are of interest but also the conditions at the liquid/substrate interface. In general, the smaller the contact angle produced by a system, the better the substrate wetting. The equipment used to measure the contact angle, theta (θ), was a VCA 2000 instrument which automatically dispensed a minute droplet of the liquid to be measured and then photographed the droplet after a set period of time, or could be programmed to take a number of photographs and measure corresponding contact angles after regular time intervals. This technique enabled the monitoring of droplets spreading on non-porous surfaces. When θ is greater than 90 degrees, as in figure 5, then beading or non-wetting is occurring, as for example shown by water on a plastic surface such as high density polyethylene. If θ is less than 90 degrees as shown in figure 5, then wetting is occurring, the smaller the angle the better the wetting process. Spontaneous wetting occurs when $\theta = 0$, the droplet immediately spreading to form a very thin continuous film on the substrate with effectively no contact angle capable of being determined.

Contact angle measurement

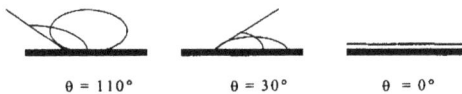

$\theta = 110°$ $\theta = 30°$ $\theta = 0°$

Figure 5 *Contact angle measurement*

Figure 6 shows the behaviour of the same surfactants on a range of low energy substrates such as polyethylene and polypropylene. For wetting it has been shown that the most efficient silicone polyether is the trisiloxane A, which has three Si atoms with pendant ethylene oxide. The low molecular weight and size gives greater mobility in solution and allows for very efficient packing at interfaces. For critical low energy substrates only the trisiloxane surfactant A allows perfect i.e. spontaneous wetting and excellent spreading.

Contact angle measurement
0.1% in water

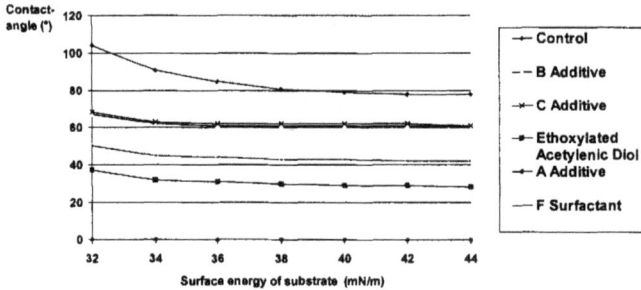

Figure 6 *Contact Angle measurement with a VCA 2000 video contact angle equipment, additive concentration at 0.1% in water*

Extending this performance behaviour now into practical life applications, the effects of these surfactants in a water-based flexographic ink formulation are shown in Figure 7. The addition of siloxane surfactant A shows excellent substrate wetting performance on plastic foil, followed by the fluorosurfactant. The siloxane surfactants B and C, which have a different structure and a higher molecular weight, do not improve the wetting behaviour. It can be seen that the acetylenic glycol has only a small impact on the wetting behaviour.

Wetting of a flexographic printing ink -
0.5% Wetting Additive

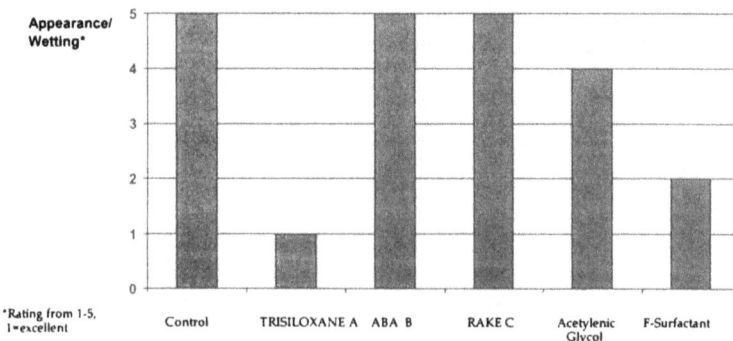

Figure 7 *Flexographic printing ink ,additive concentration 0.5%, wetting/appearance performance after application on plastic foil(surface energy 34mN/m)by draw down 12 micron, rated on a scale from 0 – 5*

Any surfactant molecule has the potential to generate foam which is undesirable in many applications. Figure 8 shows the density measurements after a high shear stir test, which give an indication of air entrapment in the ink. The siloxane surfactant C (rake) acts only as a low to moderate foam enhancer compared to the fluorosurfactant. What is also interesting is that the siloxane surfactant B, the silicone product with the highest molecular weight, is performing as a de-aerator.

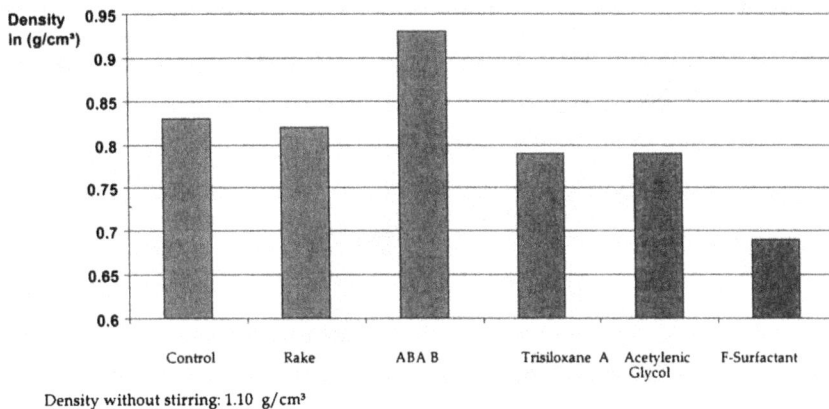

Density without stirring: 1.10 g/cm³

Figure 8 *Flexographic printing ink, additive concentration 0.5%, density measurement after dissolver stirring test (5 min at 2800 rpm) - as a measure of rate of rate foam development*

5   SILICONES AS SLIP ADDITIVES

Slip represents the lubrication of a dry coating surface. Given that the function of a coating is often to protect the underlying surface from damage whilst maintaining a satisfactory appearance, it is clear that the coating itself must be capable of resisting mechanical damage. This is where slip is important. This is often referred to as mar, rather than abrasion, resistance. In the latter, the bulk mechanical properties of the film are important, in addition to the surface lubricity.

Slip Additives must reduce friction at the coating surface. A thin layer of a material with low inter-molecular forces is capable of achieving this. The slip additive should be sufficiently compatible with the coating before application to avoid separation. After application it ought not to cause fish eyes or other defects associated with non-wetting of the additive by the coating. However, there has to be some degree of incompatibility to drive the additive to the surface of the coating film during drying.

Silicone polyethers are important as slip agents due to their structure. The ratio of the EO/PO segments has to be carefully controlled in order to achieve the required compatibility balance. Too low a polyether content may lead to the de-wetting defects already mentioned. On the other hand a high polyether content can render the copolymer too soluble, with no driving force to get it to the coating surface during drying.

As previously described for wetting, the architecture of the copolymer has a profound effect on its behaviour as a slip additive. Optimised structures have been identified by designed experimentation to give the desired combination of compatibility and slip performance. Compatibility is particularly important in clear coatings, where gloss reduction or haze are not acceptable.

Figure 9 shows slip angle results for two silicone-polyether copolymers A and B, in a water-reducible stoving paint. The main difference between the copolymers is overall molecular weight. Both products contain pendant polyether groups. It can be seen that Trisiloxane A has almost no impact on slip. This is believed to be due to the very short nature of its silicone chains. A minimum amount of dimethylsiloxy units are required to give noticeable changes in slip. This requirement is met with in rake structure B, which shows an improved slip.

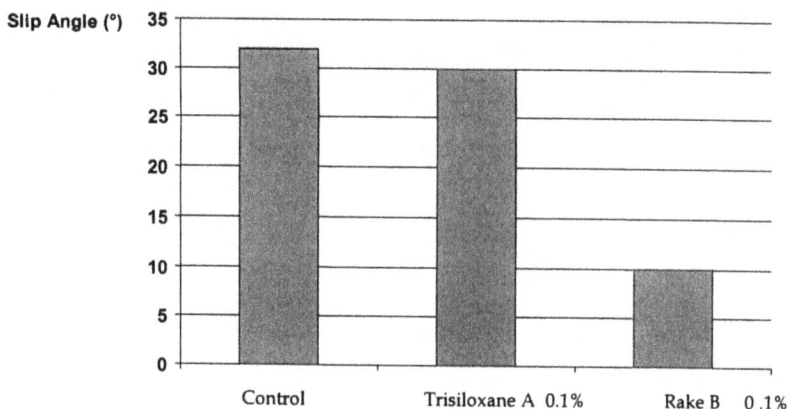

Figure 9 *Effect of molecular weight on slip performance of silicone-polyether copolymers in a water-reducible polyester stoving paint. (A - low mol. weight; B - high mol. weight)*

6   CONCLUSIONS

Silicone based surfactants offer many benefits in waterborne coatings. By the addition of the correct silicone surfactant, the coating surface tension can be modified to improve substrate wetting. In particular, the trisiloxane structure gives the best equilibrium surface tension reduction and excellent wetting to plastic surfaces and other low energy substrates. This material has the greatest capacity for lowering the liquid/solid interfacial tension. The higher molecular weight siloxane surfactants, with rake and linear structures give moderate wetting. However, materials of these types can be used to provide other benefits for instance mar resistance, slip and de-aeration.

7   REFERENCES

1. I.Schlachter and Georg Feldmann-Krane,Surfactants Sci.Ser,Silicone
      Surfactants,1998,74,201-239
2. W.Scholz,Verkroniek,10,13,1995

Subject Index

www.ingramcontent.com/pod-product-compliance
Lightning Source LLC
Chambersburg PA
CBHW032004180326
41458CB00006B/1724